Machine Learning: A Physicist Perspective

Machine Learning: A Physicist Perspective

Nelson Bolivar

ARCLER
PRESS

www.arclerpress.com

Machine Learning: A Physicist Perspective

Nelson Bolivar

Arcler Press
224 Shoreacres Road
Burlington, ON L7L 2H2
Canada
www.arclerpress.com
Email: orders@arclereducation.com

ISBN: 978-1-77469-048-2 (Hardcover)

ABOUT THE AUTHOR

Nelson Bolivar is currently a Physics Professor in the Physics Department at the Universidad Central de Venezuela, where he has been teaching since 2007. His interests include quantum field theory applied in condensed matter. He obtained his PhD in physics from the Universite de Lorraine (France) in 2014 in a joint PhD with the Universidad Central de Venezuela. His BSc in physics is from the Universidad Central de Venezuela.

TABLE OF CONTENTS

LIST OF FIGURES

LIST OF TABLES

LIST OF ABBREVIATIONS

ABC	Approximate Bayesian Computation
AGI	Artificial General Intelligence
AI	Artificial Intelligence
ANNs	Artificial Neural Networks
BDTs	Boosted Decision Trees
BMSs	Battery Management Systems
CA	Classification Accuracy
CCWC	Communication Workshop and Conference
CISIM	Computer Information Systems and Industrial Management Applications
CRFs	Conditional Random Fields
CSUR	Computing Surveys
DESSERT	Dependable Systems, Services, and Technologies
DFT	Density Functional Theory
DTs	Decision Trees
ECCV	European Conference on Computer Vision
EKF	Extended Kalman Filter
FFT	Fast Fourier Transform
GAs	Genetic Algorithms
GBDT	Gradient Boosting Decision Tree
GD	Gradient Descent
GLM	Generalized Linear Models
GN	Gauss-Newton Method
GRNN	Generalized Regression Neural Network
HCEP	Harvard Clean Energy Project
HK	Hexokinase

HMC	Hamiltonian Monte Carlo
HP	Hyperplane
ICASSP	International Conference on Acoustics, Speech, and Signal Processing
ICCCA	International Conference on Computing, Communication, and Automation
ICCNI	International Conference on Computing Networking and Informatics
ICDE'05	International Conference on Data Engineering
ICDS	International Conference on Digital Society
ICML	International Conference on Machine Learning
ICML-11	International Conference on Machine Learning-11
ICMLA	International Conference on Machine Learning and Applications
ICSD	Inorganic Crystal Structure Database
IEMECON	Industrial Automation and Electromechanical Engineering Conference
IFART	Impulse-Force-Based ART
IJERT	International Journal of Engineering Research & Technology
KNN	K-Nearest-Neighbor
KRR	Kernel Ridge Regression
LHC	Large Hadron Collider
LOOCV	Leave-One-Out Cross-Validation
LR	Logistic Regression
LRR	Linear Ridge Regression
LSST	Large Synoptic Survey Telescope
MALLET	**MA**chine **L**earning for **L**anguag**E** **T**oolkit
MAP	Maximum a Posteriori
MAPE	Mean Absolute Percent Error
MCMC	Markov Chain Monte Carlo
MGI	Materials Genome Initiative
MI	Mutual Information
MLR	Multiple Linear Regressions
MP	Materials Project
NB	Naive Bayes
NLP	Natural Language Processing
OCR	Optical Character Recognition
OQMD	Open Quantum Materials Database

PADs	Percentages of Absolute Difference
PCA	Principal Component Analysis
PGMs	Probabilistic Graphical Models
PICS	Pipeline for Images of Cosmological Strong
PLMF	Property-Labeled Materials Fragments
PLS	Partial Least Squares
PNN	Perceptron Neural Networks
PSO	Particle Swarm Optimization
QCD	Quantum Chromodynamics
QSARs	Quantitative Structure-Activity Relationship's
QSPR	Quantitative Structure-Property Relationships
RBM	Restricted Boltzmann Machine
RF	Random Forests
RL	Reinforcement Learning
RMSE	Root Mean Square Error
ROC	Receiver Operating Characteristic Curve
RUL	Remaining Useful Life
SAAs	Simulated Annealing Algorithms
SGLD	Stochastic Gradient Langevin Dynamics
SLT	Statistical Learning Theory
SOFM	Self-Organizing Feature Map
SVMs	Support Vector Machines
SVR	Support Vector Regression
TIS	Translation Initiation Site
VAEs	Variational Autoencoders
VTC	Vehicular Technology Conference
WI	Web Intelligence

PREFACE

Machine learning implies the automated and meaningful detection of patterns in the given data. In the last few decades, machine learning has been used extensively in many areas requiring the extraction of information from a large number of data sets. Machine learning has been serving the technology industry in many ways, i.e., search engines work to bring the best results for the user (while enlisting profitable ads), software secures the credit card transactions by learning how to identify frauds, and our email messages are filtered by anti-spam software. Digital cameras are programmed to detect our faces, while intelligent applications on smart devices learn to identify voice commands. Modern cars are installed with accident prevention mechanisms that are constructed using algorithms of machine learning.

Deep-learning and machine-learning have gained significant importance in the last few years. New inventions and discoveries are taking place every day to exploit the concepts of machine-learning techniques. Machine learning mechanisms are integrated with endowing programs, which have the ability to study and adapt.

The aim of this book is to present the fundamentals of machine learning with an emphasis on deep learning, neural networks, and physical aspects. The book is divided into eight chapters. Chapter 1 comprehensively discusses the fundamental concepts of machine learning and different types of machine learning methods, approaches, and algorithms. Chapter 2 deals with the physical aspects of machine learning and its integration with data science models.

Chapter 3 illustrates the concepts regarding statistical physics and machine learning. Detailed analysis of statistical models of machine learning is also provided in the chapter. Particle physics and cosmology play an important role in the development of modern machine learning technologies used in aerospace. Chapter 4 deals with the basics of particle physics and cosmology. Machine learning and artificial intelligence are playing a significant role in the modern world. Chapter 5 briefly discusses the integration of machine learning with artificial intelligence.

Material discovery and design are considered the key areas in the development of devices for commercial and everyday use. Machine learning plays an essential role in defining the characteristics and applications of materials in different areas. Chapter 6 focuses on the applications of machine learning in materials discovery and design. Presently, the field of quantum mechanics has gained significant importance due to its novel applications. Chapter 7 contains information regarding the applications of quantum physics in machine learning. Finally, Chapter 8 comprehensively discusses various applications of machine learning and its models.

The authors have made an effort to produce a self-contained and comprehensive book. However, the readers are expected to be equipped with fundamental concepts of algorithms, analysis, linear algebra, and probability. The book is aimed at explaining the fundamentals of machine learning for a general audience which includes physicists, machine learning specialists, students, and professors in the field of machine learning and quantum physics. However, readers from a machine learning and statistics background can use this book as a ready reference for their studies. Researchers can also use the advanced chapters of this book to gain a deeper theoretical insight into machine learning principles.

—Author

Chapter 1

Fundamentals of Machine Learning

CONTENTS

1.1. INTRODUCTION

Artificial intelligence (AI) has a subdivision known as machine learning. To comprehend the system of data and conjugate that data into models that can be recognized and employed by people is the general aim of machine learning (Du & Swamy, 2014; Kelleher et al., 2015).

Machine learning differs from conventional computational processes even though it is a discipline within computer science. In customary computing, calculating and deciphering problems are done by computers which are collections of explicitly automated directions called algorithms. Machine learning processes instead use arithmetic analysis to yield values that fall within a distinct range and allow for computers to train on information inputs. Consequently, to develop decision-making programs based on statistical inputs, the computer is helped by machines to construct models from test data (DasGupta, 2011; Kelleher, 2016).

Machine study has serviced anyone who uses technology today. Social media forums help users to connect and share images of friends due to facial identification technology. Images of words into a mobile type can be transformed by optical character recognition (OCR). Based on user inclination, the suggestion engines, fueled by machine learning, propose what television dramas or movies to watch. The public may soon have access to self-driving automobiles that depend on machine learning to steer (Schmid et al., 2014; Buduma & Locascio, 2017).

Machine learning is a constantly advancing field. Thus, as you analyze the magnitude of machine learning procedures and work with machine learning techniques, there are some factors to keep in view (Price et al., 2019; Yoganarasimhan, 2020).

In the following we'll look into ordinary algorithmic methodologies in machine learning, in addition to the k-nearest neighbor algorithm, in-depth learning, decision tree learning, and the standard machine learning processes of supervised and unsupervised learning. Programming languages that are most used in machine learning will be evaluated, including some of the positive and negative traits of each. Furthermore, machine learning algorithms that project certain biases will be discussed, plus when constructing algorithms that can be kept in mind to avoid these biases (Michie et al., 1994; Rasmussen, 2003).

Currently, one of the most prominent and powerful technologies is undoubtedly machine learning. A preface to the Machine Learning concepts, explaining all the basic ideas without being too high level is covered in this chapter (Holmes et al., 1994; Pedregosa et al., 2011).

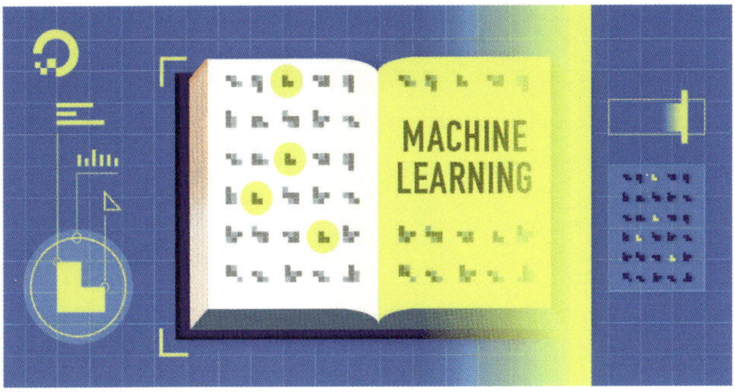

Figure 1.1. Chart for machine learning.

Source: https://www.digitalocean.com/community/tutorials/an-introduction-to-machine-learning#:~:text=2017%20258.6k-,Introduction,of%20artificial%20intelligence%20(AI).&text=Because%20of%20this%2C%20machine%20learning,has%20benefitted%20from%20machine%20learning.

Information can be translated into knowledge by a some medium, one of which is known as machine learning. There has been an outburst of data in the previous 50 years. Unless a load of data is broken down and the patterns concealed within it are found, it is futile. Invaluable underlying arrangements that we would otherwise find hard to obtain can be automatically found by machine learning means (Jordan & Mitchell, 2015). To execute all kinds of difficult decision making and foretell future events, the veiled patterns and knowledge about a dilemma can be used (Vartak et al., 2016; Herbrich, 2017).

We are drowning in information and starving for knowledge—John Naisbitt

Machine learning is connected with almost everyday affairs, yet most of us are not aware of that. It is continuously learning and enhancing from every interaction by growing a bit from the engine behind each image we take, listen to music, or google something. Creating new antidotes, determine cancer and self-driving cars are also backed up by Machine Learning (Sebastiani, 2002; King, 2009).

It is merely an inch away from all our prior rule-based systems what makes Machine Learning more entertaining:

If (x = y): do z

Typically, to produce answers to an issue, data were joined with human forged rules by software engineering. Machine Learning, on the other hand, ascertains the rules behind a dilemma by using data and answers (Chollet, 2017).

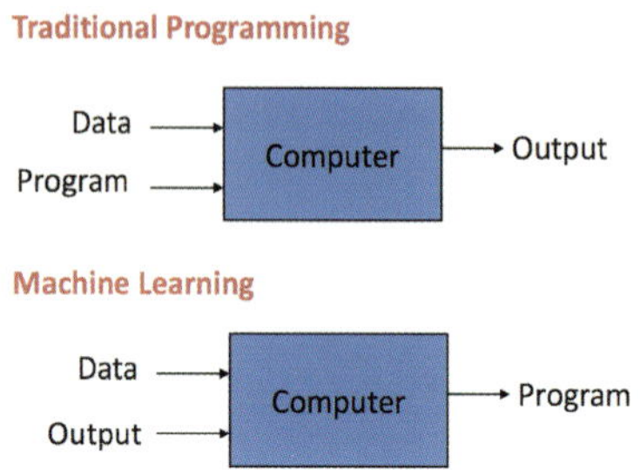

Figure 1.2. Machine learning vs. traditional programming.

[Source: https://www.pinterest.com/pin/667096382088912877/.]

Machines have to undergo a learning process to grasp from how well they execute and try distinct rules so that they can learn the directives regulating a phenomenon.

Semi-supervised, supervised, unsupervised, and reinforcement learning are several forms of Machine Learning. Although they all pursue the same fundamental theory and process, all kinds of Machine Learning have, unlike approaches. The scrutiny on each approach and the overall Machine Learning concept is included in this explanation (Dietterich, 2000; Biamonte et al., 2017).

1.2. MACHINE LEARNING: A BRIEF HISTORY

ENIAC (Electronic Numerical Integrator and Computer), the initial manually computer system was designed in the 1940s. ENIAC was named a numerical computing machine because, at that time, a person with vigorous numerical computation potentials was to be called “computer.” Improve a machine in order to be able to mimic human learning and reasoning was the primary concept since its beginning (Kononenko, 2001; Zhang et al., 2017).

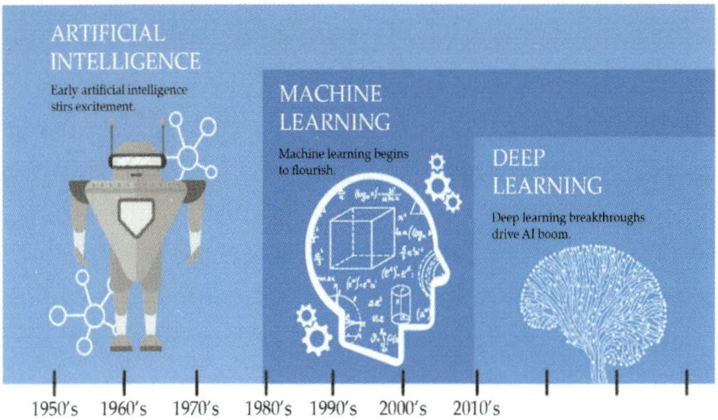

Figure 1.3. Progression of machine learning, deep learning, and artificial intelligence.

[Source: https://towardsdatascience.com/introduction-to-machine-learning-for-beginners-eed6024fdb08.]

Allegations came of being able to defeat the checker's world champion by the primal computer game program in the 1950s. This application aided checkers players a lot in improving their abilities. The Perceptron, which when integrated into grand numbers, in a matrix, became a mighty monster, but otherwise was a plain classifier that was manufactured around the same period by Frank Rosenblatt.

Figure 1.4. EIMC (electronic numerical integrator and computer).

It was a real milestone at the time as it was respective to that time. Due to its intricacy in solving particular problems, various years of inactivity of the neutral network domain were seen after that (Rouet-Leduc et al., 2017; Akbulut et al., 2018).

In the 1990s, machine learning became incredibly renowned thanks to statistics. The possible approaches in developing AI appeared, by the convergence of statistics and computer science. This deviated the field ahead toward data-driven approaches. Intelligence systems that we're able to assimilate from large amounts of information and analyze it were starting to be built by scientists due to the large-scale data available. As a headliner, the world champion of chess, the grand-master Garry Kasparov was beaten by IBM's Deep Blue system. Presently, Deep Blue is resting undisturbed in a museum following Kasparov accusing IBM of cheating but that is the past (Goswami et al., 2016; Wilson et al., 2017).

1.3. TERMINOLOGY

Below are the principal terms of Machine Learning (Navigli et al., 2003; Claveau & L'Homme, 2005; Mooney & Pejaver, 2018):

Dataset

To work out a question, an array of data examples is provided that contain significant features (Witschel, 2005; Sclano & Velardi, 2007).

Features

Machine Learning system is fed essential bits of data that teach and aids us in comprehending a problem (Song et al., 2011).

Model

When a Machine Learning algorithm recognized a fact, is an intrinsic demonstration (internal model) of it. The figures that are fed to it at the time of preparation teaches it. After the tutoring of an algorithm, this replica is the output you receive. For instance, a decision tree clone would be the production after the training of a decision tree process (Kenett & Shmueli, 2015).

1.4. MACHINE LEARNING PROCESS

Provided below is the method of the Machine Learning procedure (Doan et al., 2004; Cho et al., 2005; Ge et al., 2017):

Data Collection

Gather the facts that the algorithm will gain an understanding from.

Data Preparation

Carry out dimensionality decrement and remove important properties from the data which should be configured and set up into the optical model (Snelson, 2007).

Training

The data that has been gathered and assembled is displayed to the Machine Learning program so it learns from it and this is called the fitting stage (Vellido et al., 2012).

Evaluation

To see how finely it performs, evaluate the model.

Tuning

Optimize its efficiency by fine-tuning the design.

1.5. BACKGROUND THEORY

1.5.1. Origins

"Anything in the universe could be explained with math" It was maybe a conception claimed by the first computer analyst and one of the originators of processing, Ada Lovelace (Camastra & Vinciarelli, 2015; Sammut & Webb, 2017).

More essentially, the connection representing any event can be obtained by generating a mathematical formula. Human cooperation was not required, and machines had the aptitude to discern the world without them was discovered by Ada Lovelace (Balasubramanian et al., 2014).

The Analytical Engine weaves algebraic patterns just as the Jacquard weaves flowers and leaves

—Ada Lovelace

These foundational ideas are pivotal in Machine Learning around 200 years after. Data can be marked as data spots onto a grid no matter what the issue is. The correlation and mathematical arrangements within the prime information are tried to be found by Machine Learning algorithm (Zuccala et al., 2014).

1.5.2. Probability Theory

The probability theory which is presented into Machine Learning is build on notions that are important to it, and which were identified by another mathematician, Thomas Bayes (Muggleton & De Raedt, 1994).

Probability is orderly opinion... inference from data is nothing other than the revision of such opinion in the light of relevant new information

—Thomas Bayes

We exist in a probabilistic society. Apprehension is bound to all that happens. Machine Learning is structured upon the Bayesian expression of probability. We imagine probability as assessing the ambiguity of an affair, is what the definition of Bayesian probability argue.

Figure 1.5. The figure of probability theory.

[Source: https://towardsdatascience.com/machine-learning-an-introduction-23b84d51e6d0.]

Due to this, instead of calculating the number of recurrent trails, we have to ground our probabilities on the knowledge available regarding an event. Such as a Bayesian method would use pertinent information such as the opening team, present form, and league placing while guessing a football game instead of gauging the total number of times Manchester United has triumphed against Liverpool (Bowers et al., 1997). As the decision-making scheme is resting on logic and relevant characteristics, probabilities can still be delegated to infrequent events, which is an advantage.

1.6. MACHINE LEARNING APPROACHES

It is helpful to have previous knowledge in statistics in order to apprehend and leverage system learning algorithms as machine learning. It is an area of study which is heavily linked to computational statistics (Srinivasan & Fisher, 1995; Ng & Cardie, 2002).

To study the relationship between quantitative factors, usual practices are the correlation and regressions, whose meaning can prove to be useful for those who may not have learned statistics (Monostori et al., 1996; Shrestha & Solomatine, 2006). Two components that are not structured as either independent or dependent, hence the degree of relationship between them is called Correlation. The connection among one dependent and one independent element is studied by using Regression at an initial level. Regression allows potential prediction, because when the independent factor is known, regression statistics can be utilized to estimate the dependent variable (Guzella & Caminhas, 2009; Ye et al., 2009).

There are continuous advances in approaches to machine knowledge. Some of the famous approaches that are being applied in machine learning at the point of writing will be examined for our benefit (Nielsen et al., 1999; Lavecchia, 2015).

1.6.1. K-Nearest Neighbor

A model that can be utilized for regression, as well as categorization, is the k-nearest neighbor program which is a pattern recognition archetype. The '*k*' in the k-nearest neighbor is generally small and is a positive number, regularly abbreviated as k-NN. The k closest training samples within an area will be in the neighborhood in any of the two; classification or regression (Dudani, 1976; Keller et al., 1985).

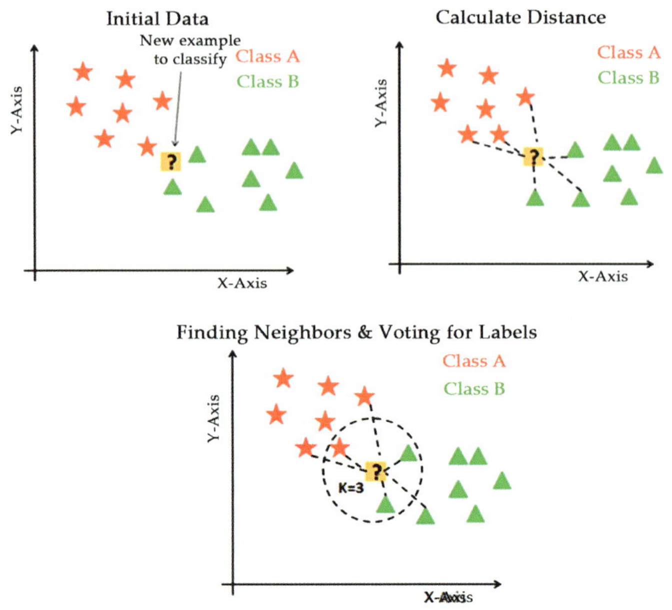

Figure 1.6. Image of K-nearest neighbor approach.

[Source: https://www.datacamp.com/community/tutorials/k-nearest-neighbor-classification-scikit-learn.]

The genus most standard amongst its k nearest neighbors will be appointed to a new object. The class of the sole closest neighbor will be designated the object in the instance of k = 1(Denoeux, 2008; Garcia et al., 2008).

As inference beyond the training information does not take place until a question is made to the network, the k-nearest neighbor is said to be a kind of "lazy learning" amongst the most principal of machine learning algorithms (Tan, 2005; Yu et al., 2005).

1.6.2. Decision Tree Learning

To show or promote decision making and to viably represent preferences, decision trees are typically employed. Decision trees are applicable as a predictive template when engaging with data mining and machine learning. The data's intended value is deduced from inspections about data by these models' outline (Freund & Mason, 1999; Kamiran et al., 2010). The aim of decision tree learning is to make a model that will foretell the worth of a target grounded on input components (Kearns & Mansour, 1999).

The outcomes about the data's desired value are presented in the leaves while the data's aspects that are specified through monitoring are represented by the sections in the predictive model.

An attribute value assessment is replicated on each of the procured subsets periodically which comprise the divided root data when "learning" a tree. The recursion procedure will be finished once the subset at a joint has a similar value as its aimed value has (Liu et al., 2008).

Whether or not one should go casting can be resolved by looking at a case of various conditions. Barometric pressure states as well as weather statuses are included in this.

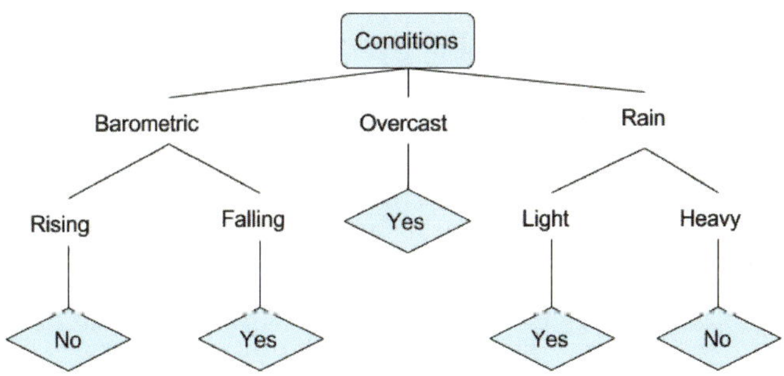

Figure 1.7. Flowchart of decision tree learning.

[Source: https://towardsdatascience.com/machine-learning-an-introduction-23b84d51e6d0.]

An example is codified by grouping it through the tree to the suitable leaf node in the clarified decision tree above. The outcome which is either a Yes or a No is then sent back the classification connected with the specific leaf. Whether a day's temperature is appropriate or not for going fishing is classified by the tree.

The interconnection should be straightforward to determine, however; a genuine classification tree data group would have many more qualities than what is summarized above. Comprehending when the decision tree has come to an evident ending, what conditions to utilize for splitting, and what characteristics to choose are various determinations that need to be made when working along with decision tree learning (Gaddam et al., 2007).

1.6.3. Deep Learning

How the human mind deals with light and auditory stimuli into sight and hearing is an endeavor at duplication by deep learning. Biological neural webbing which consists of numerous layers in a synthetic neural network built of hardware and GPUs is the influence behind deep learning architecture (Deng & Yu, 2014); LeCun et al., 2015).

To take out or convert properties (or representations) of the information, a flow of nonlinear processing unit tiers is used by deep learning. The input of a single layer acts as the output of the prior layer. Algorithms can be either unsupervised and carry out pattern evaluation or supervised and work to categorize data in deep learning.

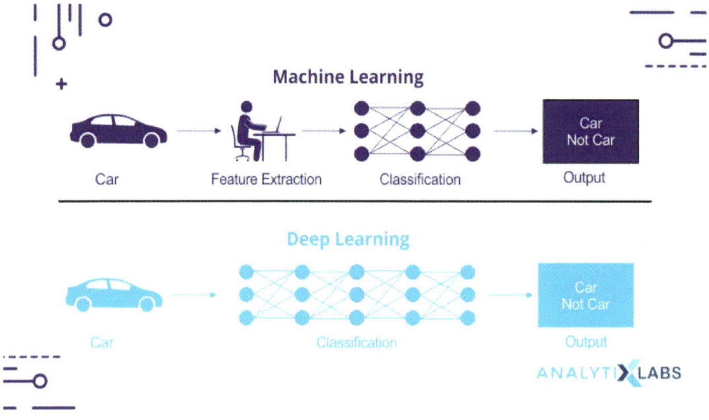

Figure 1.8. Flowchart showing deep learning vs. machine learning process.

[Source: https://www.analytixlabs.co.in/blog/how-mastery-of-deep-learning-can-trump-machine-learning-expertise/.]

Deep learning has proved capable to defeat humans in some intellectual tasks and digests the most data between the machine learning algorithms that are presently being used and enhanced. In the artificial surveillance space, deep learning has come to be the approach with considerable potential due to these qualities.

Deep learning methods have proved to realize notable breakthroughs for computer vision and dialog recognition. A prominent example of a network that exploits deep learning is IBM Watson (Chetlur et al., 2014; Schmidhuber, 2015).

1.7. MACHINE LEARNING METHODS

Functions are commonly divided into vast categories in machine learning. How response on the learning is provided to the system advanced or how learning is obtained is the footing of these categories.

Unsupervised learning which supplies the algorithm with no assigned data to enable it to search structure within its input information and, supervised learning which teaches algorithms based on sample input and output data which is tagged by humans are two of the most vastly adopted machine learning processes. These procedures will be analyzed with more scrutiny (Olden et al., 2008; Parmar et al., 2015).

While managing Machine Learning, many methods can be taken. Listed below are the zones they are generally classified into. The most frequently used and deep-rooted approaches are Supervised and unsupervised. The Learnings which have shown striking results and are fresher and more intricate are Semi-supervised and Reinforcement Learning.

The **No Free Lunch theorem** is popular in Machine Learning. Every task that you attempt to decipher has its idiosyncrasies. There is no sole algorithm that will work nicely for all processes is its statement. Hence, to fit each issue's traits, there are many algorithms and methods. Patterns that best suit various problems will keep being created of Machine Learning and AI. Below are 4 processes of machine learning (Buczak & Guven, 2015; Voyant et al., 2017):

- semi-supervised learning;
- supervised learning;
- reinforcement learning; and
- unsupervised learning.

1.7.1. Supervised Learning

Sample inputs that are denoted with their required outputs are given to the computer in supervised learning. By contrasting its real output with the "learned" outputs to find mistakes, and transform the model appropriately, the algorithm can "ascertain" and that is the aim of this approach. Accordingly, patterns are used to foresee label factors on further unlabeled figures by supervised learning (Caruana & Niculescu-Mizil, 2006; Zhu & Goldberg, 2009).

For instance, information with pictures of oceans tagged as water and shots of sharks marked as fish may be provided to the system with supervised learning. Unlabeled ocean photos as water and label-less shark images as fish should be identifiable to the supervised learning algorithm after being educated on this data.

Employing historical facts to predict statistically probably future events is a typical use type of supervised learning. Historical stock exchange information may be utilized by it to expect future fluctuations or be used to clear out spam emails. Unlabeled images of dogs can be categorized by using tagged images of dogs in supervised learning (Zhu et al., 2003; Kingma et al., 2014).

Amidst a group of inputs and outputs, the aim is to attain the mapping (the rules) in supervised learning.

For instance, the results would be the guests to the seaside and the inputs could be the weather prediction. To comprehend the mapping that explains the link between the number of beach guests and temperature is the aim of supervised learning.

The name "supervised" learning suggests that the education of the model about how it should act. Sample labeled data is given of past input and output couplets during the learning procedure. An example would be that the machine learning system will output a future forecast for the number of visitors when fresh inputs are offered in future temperatures.

The critical generalization portion of machine learning is being able to adjust to new inputs and make forecasts. The supervised model explains the authentic 'general' basic relationship because we want to increase generalization during training. The figure would be unable to conform to new, formerly unseen inputs if the model is over-educated because it ends in over-fitting to the examples used.

The guidance we supply presents bias to the learning and is a consequence to be conscious of in supervised learning. It is essential to show it trusty, unbiased samples as the design can only be echoing exactly what it was portrayed. While it comprehends, supervised learning needs a lot of information. The toughest and the costliest part of using supervised learning is oftentimes gaining enough reliably labeled data (hence, the reason why new oil is the name data is called).

For the extent of visitors to the seaside for example [high, medium, low], a section from a finite collection could be the result from a supervised Machine Learning pattern:

Input [temperature=20] -> *Model* -> Output = [visitors=high]

This is called classification as it is determining how to categorize the input in this condition.

As an alternative, a real-world scalar could be the product (output a value):

Input [temperature=20] -> *Model* -> Output = [visitors=300]

It is called regression when such is the condition.

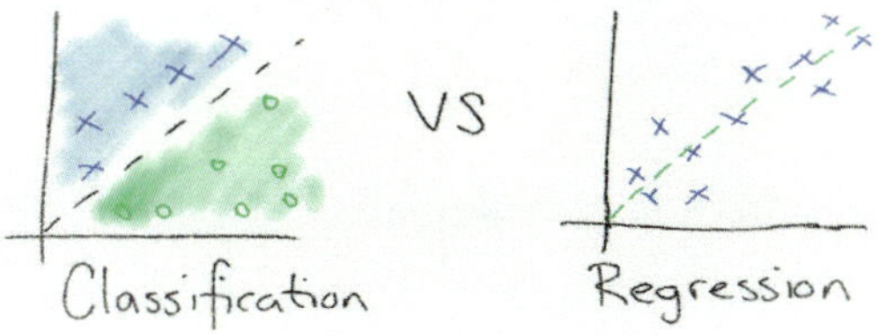

Figure 1.9. Graph for classification vs regression in supervised learning.

[Source: https://towardsdatascience.com/machine-learning-an-introduction-23b84d51e6d0.]

1.7.1.1. Classification

To categorize, the comparable data records are combined into distinct sections by something called Classification. The rules that describe how to detach the different information points are found through the use of Machine Learning.

However, how does the formation of the magical rules happen? To find the rules, various methods are available. To use information and answers to acquire rules that linearly detach data points is focused on by them (Kotsiantis et al., 2007; Ye et al., 2009).

A chief concept in machine learning is linear separability. "Can the distinctive data points be disjointed by a line?" is all that linear separability signifies. Simply put, the finest way to segregate data records with a line is tried to be found by classification approaches.

Decision boundaries is the name given to the lines depicted between classes. The decision surface is the whole area that is appointed to describe a class. A particular class will be designated to a figure point if it stands within its perimeter is what decision surface explains (Shami & Verhelst, 2007; Jain et al., 2009).

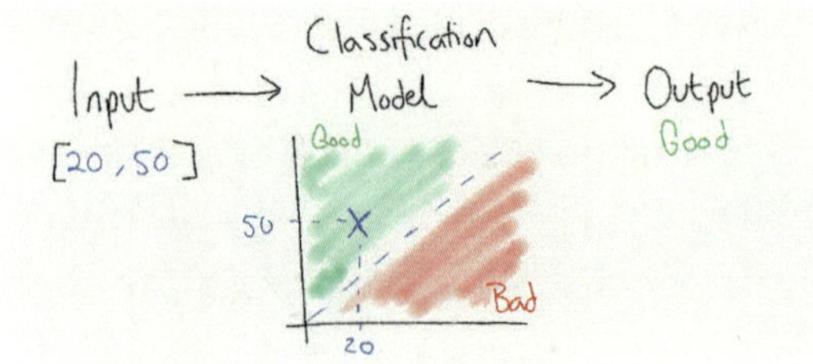

Figure 1.10. Classical model representation.

[Source: https://towardsdatascience.com/machine-learning-an-introduction-23b84d51e6d0.]

1.7.1.2. Regression

Another type of supervised learning is regression. Regression produces a number rather than a category is the distinction amidst classification and regression. Hence, when foretelling the probability of an occasion, the climate for a given day, or value-based issues like stock market costs, regression proves to be convenient (Briggs et al., 2010; Criminisi et al., 2012).

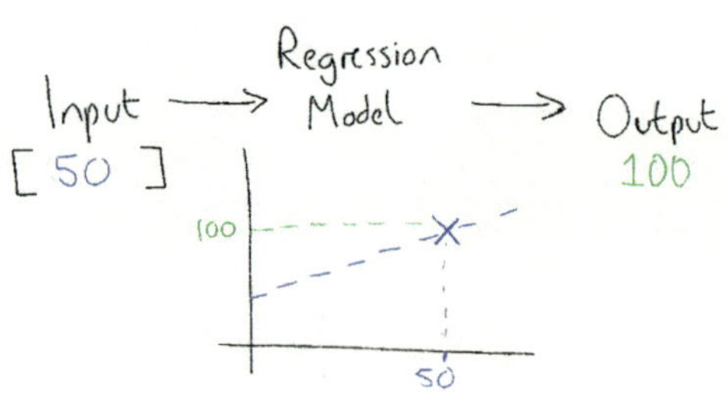

Figure 1.11. Regression model representation.

[Source: https://towardsdatascience.com/machine-learning-an-introduction-23b84d51e6d0.]

1.7.1.3. Examples

To choose when to purchase/sell and make a profit, regression is used in fiscal trading to search the designs in stocks and other goods. If an email you get is spam, classification is already being used to classify it. Supervised learning methods: regression and classification can be widened towards much more intricate tasks. For instance, oration and sound tasks. Some examples are object spotting, chatbots, and picture classification.

A model educated with supervised learning to practically fake videos of humans talking is used in (Dasgupta et al., 2011). How does classification or regression have a concern with this complicated image-based task, you think? Everything in the universe is explained with math and numbers, even intricate phenomenon is what it comes back to. Similar to regression, a neural system is still only producing values in this example. However, numerical 3d coordinate numbers of a facial net are the figures in this instance.

1.7.2. Unsupervised Learning

The learning system is left to search similarities among its input information as data is unlabeled in unsupervised learning. Machine learning techniques that help unsupervised learning are especially valuable because unlabeled data are more extensive than labeled data (Figueiredo & Jain, 2002).

Unsupervised learning may have an aim of feature studying through which the outlines that are required to categorize raw information are reflexively obtained by the computational machine, but it may also be as direct as finding hidden designs within a dataset.

Transactional figures are what unsupervised learning is generally used for. To work out what alike qualities can be taken out from customer descriptions and their kinds of purchases from a huge dataset of clients and their buying may be improbable for you as a person. A marketing operation associated with pregnancy and baby merchandise can be aimed at the clientele to maximize their number of purchases after calculating those women of a particular age range who purchase unscented soaps are more probable to be pregnant by entering this information into an unsupervised learning system (Hofmann, 2001).

Difficult data that is wider and unrelated can be structured in potentially coherent ways by unsupervised learning techniques without being informed of a "right" answer. Recommender networks that suggest what goods to purchase next, oddity detection, and for crooked credit card buying is what unsupervised learning is mostly used for. To search similarities and categorize dog images together, untagged pictures of dogs can be utilized as input information for the system in unsupervised learning.

In the examples, only input statistics are given in unsupervised learning. There are no labeled instance productions to aim for. Surprisingly, many appealing and labyrinthine designs hidden within the information without any notes are still likely to be found. Sorting various color coins into distinct piles is a real-life instance of unsupervised learning. You can recognize which color coins are linked and sort them into their accurate groups by just seeing their attributes such as color even though no one taught you how to segregate them.

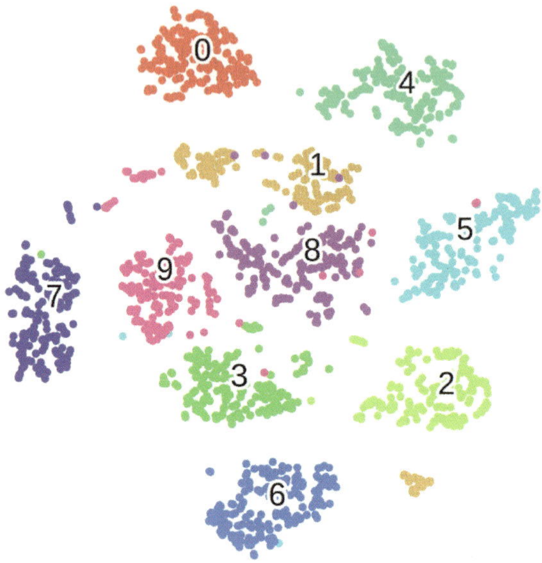

Figure 1.12. Rooted only on their features, handwritten figures are accurately sorted into groups by an unsupervised learning algorithm (t-SNE).

[Source: https://www.oreilly.com/content/an-illustrated-introduction-to-the-t-sne-algorithm/.]

The issue becomes less defined with the expulsion of supervision; hence, unsupervised learning can be tougher than supervised learning. A concept of what designs to look for is less focused on the system.

Imagine during your studying sessions. By re-using the supervised information of notes, harmony, and rhythms, you would comprehend how to play guitar faster under the supervision of an instructor. However, you would find it so much more difficult knowing where to begin if you only taught yourself.

You begin from a clean slate with less partiality and may even find a fresh, better way to decipher an issue by being unsupervised in a laissez-faire educating style. Thus, that is the reason knowledge history is what unsupervised learning is also called. When performing an exploratory information analysis, unsupervised learning is quite necessary.

To find the interesting structures in unlabeled data, we use density estimation. The most common form of which is clustering. Among others, there is also dimensionality reduction, latent variable models, and anomaly detection. More complex unsupervised techniques involve neural networks like Auto-encoders and Deep Belief Networks, but we will not go into them in this introduction blog.

We use density estimation to look for the appealing structures in unlabeled figures. Clustering is the most general form. Dormant variable models, anomaly recognition, and dimensionality reduction amidst others are also within them. Neural systems like Deep Belief Networks and Auto-encoders are engaged in more complicated unsupervised methods, but we will not get into them in this chapter.

1.7.2.1. Clustering

Clustering is for what unsupervised learning is applied. Creating sections with differing features is the act called Clustering. Within a data file, several subgroups are strived to be found by Clustering. We are not limited to any group of labels and are at liberty to pick how many clusters to make as this is unsupervised learning. This is both a burden and a fortune. An empirical design selection procedure is performed after choosing a model that has the appropriate number of clusters (complexity) (Chen et al., 2005; Caron et al., 2018).

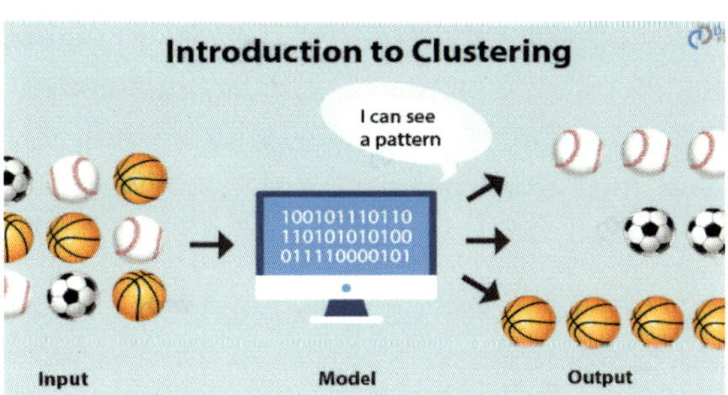

Figure 1.13. Machine learning's application; data clustering.

[Source: https://data-flair.training/blogs/clustering-in-machine-learning/.]

1.7.2.2. Association

You want to reveal the rules that explain your data in Association Learning. If someone watches video A, they will probably watch video B is an instance of it. For examples such as this where you wish to look for related things, association directions are perfect (Cios et al., 2007; Li et al., 2018).

1.7.2.3. Anomaly Detection

The recognition of odd or uncommon items that diverge from the mass of data. For instance, fraudulent movement on your card will be noticed through the application of this by your bank. Within a standard range of attitudes and values is where your standard spending routines will fall into. The attitude will be distinct from your normal habit, however, when anyone attempts to rob you using your card. To detach and recognize these weird happenings, anomaly detection uses unsupervised learning (Schlegl et al., 2017).

1.7.2.4. Dimensionality Reduction

An inferior more systematic set that still encrypts the significant data after the alleviation from the initial feature set down is the most crucial quality aimed to be found by Dimensionality reduction.

For instance, the day of the week, month, the number of occasions arranged for that day, and the weather might be used as inputs while foretelling the number of guests to the seaside. For forecasting the number of guests, however; the month might not be essential (Kumar et al., 2017; 2019).

Less precision and effectiveness are created by unimportant characteristics such as this that make a Machine Learning system confused. Only the most crucial qualities are recognized and used by using dimensionality reduction. A generally used method is Principal Component Analysis (PCA).

1.7.2.5. Examples

Grounded on the star's features, a new kind of star by inspecting what subsets of star automatically manifest has been founded by using clustering. To divide customers into alike groups based on their attitudes and qualities, it is commonly used in marketing.

To suggest or look for related things, associating learning is used. Market basket analysis is a general example. Based on what they have put in their cart, association conditions are found to foretell other things a customer is expected to purchase in market basket analysis. Amazon utilizes this. They suggest things like a laptop case through their association rules if you put a new laptop in your cart.

Situations such as fraud recognition and malware recognition, anomaly detection are compatible.

1.7.3. Semi-Supervised Learning

A blend amongst supervised and unsupervised approaches is called semi-supervised learning. Semi-supervised learning takes the central path. We do not let the system do its own thing and give no type of feedback, but the learning procedure is not carefully supervised with example productions for every single input (Zhu et al., 2005; Zhu & Goldberg, 2009).

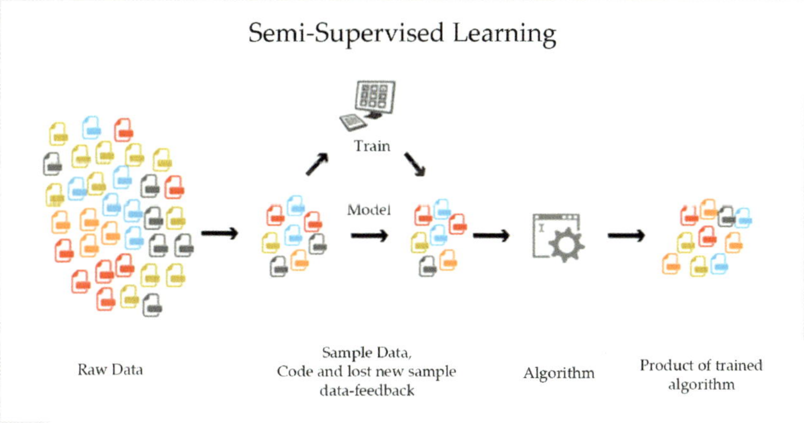

Figure 1.14. Semi-supervised machine learning illustration.

[Source: http://primo.ai/index.php?title=Semi-Supervised.]

To lessen the weight of having adequate labeled information, a much bigger unlabeled dataset is combined with less amount of labeled information. Hence, Machine Learning's capability to resolve many more issues is opened by it.

1.7.3.1. Generative Adversarial Networks

A current advancement with amazing outputs has been Generative Adversarial Networks (GANs). A b are two neural systems used by GANs. The discriminator reviews the product, and the generator creates it. They both become progressively proficient by competing against each other.

It can be categorized as semi-supervised because by utilizing a system to both create input and another one to generate the result, there is no need for us to give obvious labels every single time (Zhang et al., 2019).

1.7.3.2. Examples

Breast cancer examinations are an excellent example of medical tests. It is laborious and very costly, and a skilled expert is needed to tag these. The semi-supervised system would be able to mark a larger group of scans if a professional can label just a slight set of breast cancer tests by using the small subset.

The most splendid instances of semi-supervised learning, for me, are GAN's. Unsupervised learning is being utilized to connect aspects from a picture to the other by a Generative Adversarial Network in the video below.

Without using designated training information, a neural system known as a GAN (generative adversarial network) is used to fuse pictures.

1.7.4. Reinforcement Learning

You will have noted reinforcement learning if you are accustomed to psychology. We study it in daily life, so you will already know the basis if not. To strengthen behaviors in this method, a random positive and negative response is used. Good behaviors are compensated with a present and become more ordinary just like if imagine it as if instructing a dog. Bad behaviors are chastened and become less ordinary. In reinforcement learning, this present-motivated attitude is key (Kaelbling et al., 1996; Mnih et al., 2015).

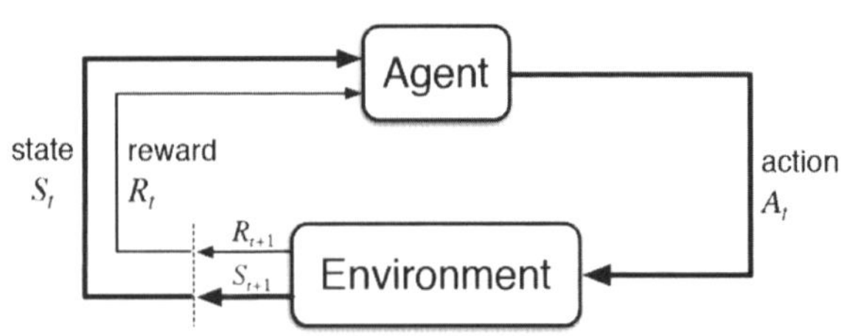

Figure 1.15. Reinforcement learning being influenced by cyclic.

[Source: kdnuggets.com/2018/03/5-things-reinforcement-learning.html.]

The style we as humans learn is quite alike to it. We frequently learn by getting positive and negative indications throughout our lives. We get these gestures by the substances in our brain which is one of many methods. We become more probable to redo that particular action which makes us feel pleasant because the neurons in our minds give a boost of positive neurotransmitters such as dopamine when anything nice happens. Like in supervised learning, we do not need frequent supervision to understand. We still learn very successfully by only providing the incidental reinforcement signals.

It is an initial step away from teaching on static datasets, and rather than being able to employ dynamic, loud data-rich environments is one of the most thrilling parts of Reinforcement Learning. A learning method used by humans is brought nearer to Machine Learning. The Earth is simply our loud, intricate data-rich environment.

In Reinforcement Learning investigation, games are very renowned. Classic data-rich surroundings are delivered by them. To teach reward-motivated attitudes, the points in games are fitting reward signals. Plus, to decrease overall teaching time, time can be accelerated in an artificial game environment.

By playing the game repeatedly, a Reinforcement Learning system just plans to increase its rewards. It is expected to be fitted to Reinforcement Learning if you can mount an issue with a constant 'score' as a gift.

Because of how fresh and complicated it is, reinforcement learning has not been employed as much in the actual world. However, to decrease information center running prices by directing the cooling networks more productively, the use of reinforcement learning is an actual-world example. To receive the lowest energy prices, the system learns an optimal strategy of how to act. The more gifts it gets is linked to how low the price is.

It is constantly used in games in testing. Imperfect data (where sections of the state are concealed, e.g., the actual world) and perfect data (where you can see the whole state of the surrounding) in games have both seen an extraordinary success that eclipse humans.

1.8. MACHINE LEARNING ALGORITHMS

In the extent of supervised learning where which mostly classification is dealt with, following are the types of algorithms:

- Linear Classifiers
- K-Means Clustering
- Quadratic Classifiers

- Perceptron
- Random Forest
- Boosting
- Naïve Bayes Classifier
- Logical Regression
- Support Vector Machine
- Decision Tree
- Bayesian Networks
- Neural Networks

A few algorithms are discussed in detail in following sections.

1.8.1. Linear Classifiers

In the process machine learning, classification is used to group objects with comparable feature values, into the groups. Timothy et al. (1998) declared that the linear classifiers accomplish this with the help of a classification decision (Grzymala-Busse & Hu, 2000; Grzymala-Busse et al., 2005). The classification decision is dependent on the input of linear combination. If the real vector $\vec{x}$ is the input to the classifier, then the output is given as:

$$y = f\left(\vec{w}.\vec{x}\right) = f\left(\sum_j w_j x_j\right)$$

Where, $\vec{w}.\rightarrow$ real weights vector, f→ function in which the dot product of two vectors is translated into the preferred output. A set of marked training samples helps to deduce the vector $\vec{w}$.. Usually, the values above a specific threshold are mapped to first class and the rest of the values to second class by the function f. The probability that a certain item fits in a certain class is given by the complex function f (Li et al., 2004; Honghai et al., 2005; Luengo et al., 2012).

In the two-class classification, the functionality of the linear classifier can be visualized as dividing with hyperplane the high-dimensional input space. The points on the one side of hyperplane are marked as "yes," while the points on the other side are marked as "no." Linear classifiers are frequently used in conditions where the pace of classification is a problem because linear classifiers are the fastest classifiers especially when the real vector $\vec{x}$ is sparse. The decision trees are also capable to be faster (Hornik et al., 1989; Schaffalitzky & Zisserman, 2004). Linear classifiers usually perform very well when in the real vector $\vec{x}$ the number of dimensions is large. In document classification, every element in the real vector $\vec{x}$ is usually the counts of a certain word in the document. The classifier must be well-regularized in these cases (Dempster et al., 1997; Haykin & Network, 2004).

SVM models are closely related to the traditional multilayer (PNN) perceptron neural networks. Using a kernel function, Support vector machine models are the substitute training scheme for the polynomials, radial basis functions and multi-layer perceptron neural networks classifiers. In SVM the weight of the network is found by resolving the quadratic programming problem having linear constraints, instead of resolving a non-convex and unconstrained minimization problem such as used in the standard neural network teaching (Gross et al., 1969; Zoltan-Csaba et al., 2011; Olga et al., 2015).

As indicated by Luis- Gonz (2005), support vector machine executes classification by creating the *N*-dimensional hyperplane which optimally splits the data into categories. Support vector machine models are intricately connected to neural networks. The Support vector machine model, in which the sigmoid kernel function is used resembles the two-layer (PNN) perceptron neural network (Rosenblatt, 1961; Dutra da Silva et al., 2011).

In the language of support vector machine literature, the *attribute* is a predictor variable, and an altered attribute which defines the (HP) hyperplane is known as a *feature*. The process of selecting the most appropriate representation is called the feature *selection*. Set of features which describes one of the cases is known as a *vector*. The goal of SVM (support vector machine) modeling is to discover the optimum hyperplane which splits the groups of vectors in a way that cases with one category of the target variable are on one side of the plane and cases with the other category are on the other side of the plane (Block et al., 1962; Novikoff, 1963; Freund & Schapire, 1999). The vectors closest to the hyperplane are known as the *support vectors*.

1.8.2. K-Means Clustering

In the beginning K (number of the cluster) is determined and the center of the clusters is assumed. Any random objects are taken as the primary center or the first K objects in the structure can also function as the primary center (Teather, 2006; Yusupov, 2007).

Then three steps mentioned below are performed by the K means algorithm until convergence. Iterate until stable.

- Determining the distance of every object from the center.
- Determining the center coordinate.
- Grouping the objects based on the minimum distance.

K-means flowchart is shown in the Figure 1.16.

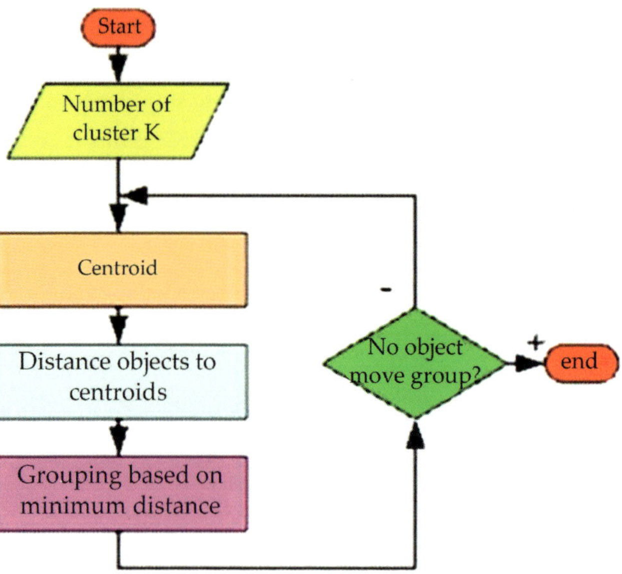

Figure 1.16. Schematic illustration of K-means iteration.

[Source: https://www.intechopen.com/books/new-advances-in-machine-learning/types-of-machine-learning-algorithms.]

K-means clustering is an easiest unsupervised learning algorithm which provides the solution of the well-known clustering problem. The process adopts a simple way to categorize the given data set amongst the definite number of clusters. For each cluster, K centroid is to be defined. These centroids must be positioned in a scheming way because different results can be obtained from a different location (Franc & Hlaváč, 2005; González et al., 2006). The more appropriate choice is to position the centroids as far as possible from each other. The next step involves taking each point from the given data and associating the point to the adjacent centroid. When there is no point left, it indicates the completion of the first step and an early grouping is done (Oltean & Dumitrescu, 2004; Fukunaga, 2008; Chali,

2009). Now here re-calculation of 'k' new centroids is to be done. After obtaining the new centroids, a fresh binding amongst the same points and the adjacent new centroid is to be done. A loop is produced. Due to this loop, the k centroids adjust their position gradually until there are no more changes. Thus the centroids don't move anymore (Oltean & Groşan, 2003; Oltean, 2007). This algorithm goal at diminishing an *objective function*, a squared error function in this particular case. The objective function:

$$J = \sum_{J-1}^{k} \sum_{i-1}^{n} \left\| x_i^{(j)} - c_j \right\|^2 ,$$

$\left\| x_i^{(j)} - c_j \right\|^2$ Is a selected distance measure between $x_i^{(j)}$ (data point) and c_j (cluster center). It is a sign of distance of *n* data points with respect to their particular cluster centers.

The algorithm for this technique consists of the steps defined below:

- K points are placed into the space denoted by objects which are being clustered. These points denote primary group centroids.
- Each object is assigned to the group which has the nearest centroid.
- The positions of K centroids are recalculated after all of the objects have been allocated.
- Steps ii and iii are repeated as long as the centroids are moving. This creates a partition of objects into the groups from where the minimized metric can be calculated.

Though it can be ascertained that the process will always terminate, this k-means clustering algorithm does not essentially find the most optimum configuration equivalent to the minimum global objective function (Burke et al., 2006; Bader-El-Den & Poli, 2007). The K-means algorithm is also considerably delicate to the primary arbitrarily selected cluster centers. This algorithm is run numerous times to diminish this effect. K-means algorithm has been modified to many problem fields. This algorithm is a worthy contender for modification to work with the fuzzy feature vectors (Tavares et al., 2004; Keller & Poli, 2007).

Assume that n-sample feature vectors (x_1, x_2, x_3 ..., x_n) are available, all of the similar class and they plunge into the k dense clusters where k < n. Assume m_i to be the mean in the cluster i. We can implement the minimum distance classifier to split the clusters if they are well separated. It can be said that the x is present in the cluster I only if $\| x - m_i \|$ represents the minimum of all k distances. This recommends the following process in order to find the k means:

Supposing the values for means (m_1, m_2, m_3 ..., m_k) unless no modifications in any of the mean occur

- Classifying the samples into the clusters using the estimated means
- For the cluster i from the range 1 to k
- Swap m_i by, mean of all the samples for the cluster i
- end_for
- end_until

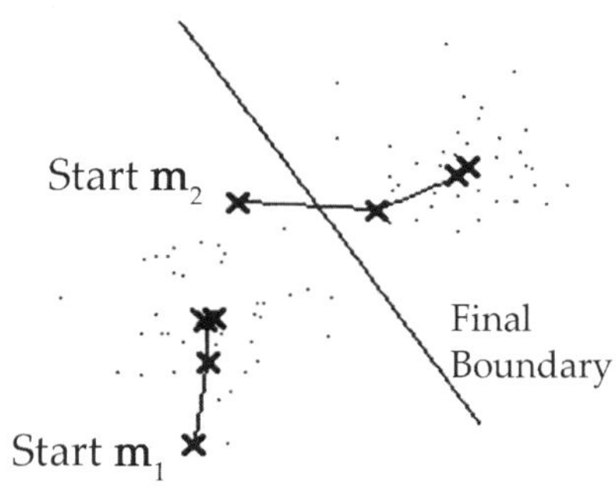

Figure 1.17. Demonstration of the motion for m1 and m2 means at the midpoint of two clusters.

[Source: https://www.intechopen.com/books/new-advances-in-machine-learning/types-of-machine-learning-algorithms.]

This is an easy form of the k-means process. It can be observed as the greedy algorithm for separating the n models into k clusters to reduce the totality of squared distances to cluster centers. Some weaknesses of this algorithm do exist:

- The method to set the means wasn't stated. One famous way to start the mean is to arbitrarily choose k of the models.
- It is possible that a set of models nearby to m_i is vacant so that the m_i can't be upgraded. This is an irritation which must be looked after in an implementation.
- The outcomes yielded are dependent on the primary values for the mean, and it commonly happens that the suboptimum partitions

are originated. The standard solution for this is to go for a number of dissimilar starting points.

- The outcomes are dependent on the metric which is used to quantify $|| x - m_i ||$. The standard way out is to standardize every variable by its typical deviation, although this is not desirable every time.

1.8.3. Neural Network

Neural networks can, in fact, execute a number of reversion and classification jobs at once, though normally each network executes only one (Forsyth, 1990; Bishop C. M., 1995; Poli et al., 2007). In the majority of cases, there is only one particular output variable for the network. For many state classification problems, this may relate to the number of outcome units. If you outline a network with numerous output variables, then it might suffer from cross-talk. The concealed neurons have difficulty in learning as they are trying to model as a minimum two functions. The best possible solution is that for each output, train the separate network and then combine them into one so that they can perform as a unit. Neural network methodologies are listed below:

- multilayer perceptron;
- training multilayer perceptron;
- back propagation algorithm;
- over-learning and generalization; and
- selection of data.

1.8.4. Self-Organized Map

SOFM (Self-Organizing Feature Map) networks are utilized in a different way as compared to other networks. Other networks are planned for the supervised learning tasks whereas the Self-Organizing Feature Map networks are designed mainly for the unsupervised learning tasks (Haykin, 1994; Jain et al., 1996; Fausett, 1994). Training data set in supervised learning comprises of the cases containing input variables organized with the related outputs whereas the training set in unsupervised learning consists of input variables only. This may appear weird at the first glance. What can a network learn without the outputs? The answer to this question is the Self-Organizing Feature Map network tries to absorb the configuration of the data. Kohonen (Kohonen, 1997) explained that one possible use of SOFM is in the exploratory data examination. The Self-Organizing Feature Map (SOFM) network can pick up to identify the data clusters and can relate classes which are alike. The consumer of the network can figure out the understanding of data, which is then utilized to improve the network. As the classes of the data are acknowledged, they can then be labeled in order to make the network capable of the classification tasks. The Self-Organizing Feature Map networks can be used for classification when the output classes are available instantaneously. The benefit in this situation is their capability to highlight the resemblances between the classes.

A second potential use of this network is in the novelty detection. The Self-Organizing Feature Map networks can acquire to identify the clusters in training data and then respond. If the new data come upon, the network is unsuccessful to identify it and this specifies novelty.

The Self-Organizing Feature Map network consists of the following two layers:

- An output layer with radial units (topological map layer). In topological map layer, the units are positioned in space, i.e., in the two dimensions.
- The input layer.

With the help, an iterative algorithm the Self-Organizing Feature Map networks are skilled (Jain et al., 1996). Beginning with the random set of the radial centers, the algorithm adjusts them progressively to reveal the clustering of the data. This algorithm, at one stage, relates with sub-sampling and the *K*-Means algorithms which are used to allocate centers in the SOM (Self-Organize Map) network. The Self-Organizing Feature Map algorithm can be utilized to allocate centers for these kinds of networks.

The basic iterative SOFM algorithm goes through a number of periods, on every single epoch implementing the training case and employing the following algorithm:

- Selection of the winning neuron. The neuron whose center is closest to the input situation is known as the winning neuron.
- Adjusting the winning neuron so that it can resemble the input case.

The iterative training technique also organizes the network in a way that the units demonstrating centers close to each other in input layer are also located close to each other on the topological layer. The topological layer of the network can be thought of as a crude 2-dimensional lattice, which must be crumpled and slanted into N-dimensional input layer to preserve the original structure in the best possible way. Any effort to characterize the N-dimensional space into the two dimensions will outcome in the loss of detail. This method can be useful in letting the user think about the data which otherwise might be difficult to understand.

The iterative algorithm utilizes the time decaying learning rate, implemented to accomplish the weighted sum. It also confirms that the modifications become more delicate as the period's pass. This guarantees that the centers snuggle down to a cooperative demonstration of the cases that helps the particular neuron to win. By adding the thought of neighborhood to the iterative algorithm the topological assembling property is accomplished. The set of neurons adjacent to the winning neuron is called a neighborhood. The neighborhood similar to the learning rate deteriorates over time. Initially, a big number of neurons are located in the neighborhood whereas with the passage of time the neighborhood will be empty. In the Kohonen or SOFM algorithm, the modification of neurons is practiced on all the fellows of the current neighborhood.

The consequence of this neighborhood apprises is that primarily big areas of a network are pulled towards the training cases and are dragged considerably. The network cultivates a crude topological arrangement with the similar cases triggering the clusters of neurons inside the topological map. Both the learning rate and the neighborhood decreases with the passage of time so that the better differences within the areas of the map can be made, eventually resulting in the fine-tuning of an individual neuron. Very often the training is intentionally organized in two separate phases:

- A comparatively short phase with usually the high learning rates and neighborhood.
- A long phase having the low learning rates and zero neighborhoods.

Self-Organizing Feature Map networks, when executing classification also utilizes the accept threshold. The triggering extent of the neuron in the Self-Organizing Feature Map network is its distance from the input, the accept threshold plays the role of the maximum documented distance. If the initiation of winning neuron is much larger as compared to this distance then the Self-Organizing Feature Map network is considered as undecided. By tagging all the neurons and allocating the accept threshold suitably, a Self-Organizing Feature Map network can perform as the novelty detector.

Self-Organizing Feature Map network as articulated by Kohonen (Kohonen, 1997) are motivated by some of the known characteristics of the brain. The cerebral cortex is really a large smooth sheet with identified topological properties.

Once the network is able to identify configuration in the data, the network can be utilized as a visualization device to inspect the data. With the help of this network Win, Frequencies Datasheet can be observed to see if the discrete clusters have been made on the map. The Individual cases are implemented and topological map examined, to observe if some significance can be allotted to clusters. Once the clusters are recognized, the neurons present in the topological chart are labeled to specify their meaning. When the topological chart has been plotted in this way, then new circumstances can be given to network. The network can execute arrangement on the condition that winning neuron must be tagged with the class name. If not, then the network is considered as undecided.

1.9. PROGRAMMING LANGUAGES

For the employment in machine learning procedures, you may want to contemplate libraries available in multiple languages coupled with the abilities listed on present job advertisements when selecting a language to master in with machine learning.

Python is the most asked-for computer language in the machine learning professional discipline inferred from information taken from job advertisements on indeed.com in December 2016. Java, then R, then C++ comes behind Python.

The maximized development of deep learning structures available for this language presently, including TensorFlow, PyTorch, and Keras may be linked to Python's acclaim. Python confirms to be mighty and direct both for preprocessing information and working with data precisely as a language that has legible syntax and the quality to be utilized as a scripting language. NumPy, SciPy, and Matplotlib are various existing Python items that Python developers may already be accustomed to and the scikit-learn machine learning library is made on top of it.

You can read our workshop series on "How To Build a Machine Learning Classifier in Python with scikit-learn" or "How To Perform Neural Style Transfer with Python 3 and PyTorch" or "How To Code in Python 3" to begin with Python.

Apex desktop application designers who are also grappling with machine learning at the campaign level commonly use Java which is extensively used in enterprise programming. It is preferred by those with experience in Java advancement to pertain to machine learning and generally, it is not the initial choice for those fresh to programming who want to study machine learning. Alongside cyber-attack and fraud recognition use tasks, Java tends to be employed more than Python for system security in terms of machine education applications in the industry.

MALLET (**MA**chine **L**earning for **L**anguag**E** **T**oolkit) permits machine learning applications on text, in addition to native language processing, topic designing, document categorization, and clustering; and Weka, a set of machine learning systems to use for information mining cases and Deeplearning4j, an open-source and divided deep-learning assemblage written for both Scala and Java are amid machine learning libraries for Java.

For statistical processing, an open-source coding language used chiefly for it is **R**. It is preferred by many in academic circles and in recent times, has grown in fame. Due to an elevated interest in information science, it has soared in industrial applications but it is not commonly used in industrial manufacturing environments. e1071 which comprises tasks for statistics and probability philosophy, caret (which is the abbreviation of **C**lassification **A**nd **RE**gression **T**raining) for making predictive designs and general Forest for categorization and regression are contained in renowned packages for machine learning in R.

For artificial intelligence and machine learning, **C++** is the language of desire in-game or robot software (inclusive of robot locomotion). Because of their expertise and amount of command in the language, installed computing hardware designers and electronics engineers are more inclined to prefer C++ or C in machine learning programs. The open-source and modular Shark, scalable ml pack, and Dlib offering far-reaching machine learning algorithms are a few machine learning libraries you can utilize with C++.

1.10. HUMAN BIASES

Being built on the information does not mean that machine learning products are neutral, that is certainly not the case; even if data and computational examination may make us think that we are getting objective information. How machine learning will work with that data heavily depends upon human prejudice and the role it plays in how information is collected, arranged, and ultimately in the algorithms that decide the interaction among data and machine learning (Char et al., 2018).

A computer may not categorize a shark as a fish in the instance when individuals are supplying pictures for "fish" as information to teach an algorithm, and these individuals overwhelmingly choose images of goldfish. Sharks would not be considered as fish as this would develop a bias against sharks as fish.

A computer may not appropriately categorize scientists who are also humans of color or women when using historical images of scientists as training information. AI and machine learning software display human-like biases that consist of race and gender intolerance according to the indication of fresh peer-reviewed research.

Systematic problems prolonged by uncaught biases may stop people from getting same-day delivery options, from being authorized for loans, or from being presented ads for high-paying job possibilities as machine learning is growingly leveraged in business (Caliskan et al., 2017).

It is very important to be mindful of human bias and to work towards removing it as much as imaginable as it can negatively affect others. Making sure that diverse individuals are testing and critiquing a project and diverse individuals working on a scheme is one way to work towards obtaining this. To perform ethics analysis as part of information science project preparation, building substitute systems that can recognize biases, and to observe and audit algorithms, regulatory third parties have been reached out for by others. To fight bias in this field, it is essential to be conscious of our own unconscious biases, organizing equity in our machine learning programs and pipelines, and boosting awareness about biases.

REFERENCES

1. Akbulut, A., Ertugrul, E., & Topcu, V. (2018). Fetal health status prediction based on maternal clinical history using machine learning techniques. *Computer Methods and Programs in Biomedicine, 163*, 87–100.
2. Bader-El-Den, M., & Poli, R. (2007, October). Generating SAT local-search heuristics using a GP hyper-heuristic framework. In *International Conference on Artificial Evolution (Evolution Artificielle)* (pp. 37–49). Springer, Berlin, Heidelberg.
3. Balasubramanian, V., Ho, S. S., & Vovk, V. (Eds.). (2014). *Conformal Prediction for Reliable Machine Learning: Theory, Adaptations, and Applications* (Vol.1, pp. 1–24). Newnes.
4. Biamonte, J., Wittek, P., Pancotti, N., Rebentrost, P., Wiebe, N., & Lloyd, S. (2017). Quantum machine learning. *Nature*, *549*(7671), 195–202.
5. Bishop, C. M. (1995). *Neural Networks for Pattern Recognition.* Oxford, England: Oxford University Press.
6. Block, H. D., Knight Jr, B. W., & Rosenblatt, F. (1962). Analysis of a four-layer series-coupled perceptron. II. *Reviews of Modern Physics*, *34*(1), 135.
7. Bowers, A. F., Giraud-Carrier, C., Kennedy, C., Lloyd, J. W., & MacKinney-Romero, R. (1997, September). A framework for higher-order inductive machine learning. In *Proceedings of the COMPULOGNet Area Meeting on Representation Issues in Reasoning and Learning* (Vol.1, pp. 19–25).
8. Briggs, F. B. S., Ramsay, P. P., Madden, E., Norris, J. M., Holers, V. M., Mikuls, T. R., ... & Barcellos, L. F. (2010). Supervised machine learning and logistic regression identify novel epistatic risk factors with PTPN22 for rheumatoid arthritis. *Genes & Immunity*, *11*(3), 199–208.
9. Buczak, A. L., & Guven, E. (2015). A survey of data mining and machine learning methods for cyber security intrusion detection. *IEEE Communications Surveys & Tutorials*, *18*(2), 1153–1176.
10. Buduma, N., & Locascio, N. (2017). *Fundamentals of Deep Learning: Designing Next-Generation Machine Intelligence Algorithms.* "O'Reilly Media, Inc." (Vol.1, pp. 1–33).

11. Burke, E. K., Hyde, M. R., & Kendall, G. (2006). Evolving bin packing heuristics with genetic programming. In *Parallel Problem Solving from Nature-PPSN IX* (pp. 860–869). Springer, Berlin, Heidelberg.
12. Caliskan, A., Bryson, J. J., & Narayanan, A. (2017). Semantics derived automatically from language corpora contain human-like biases. *Science, 356*(6334), 183–186.
13. Camastra, F., & Vinciarelli, A. (2015). *Machine Learning for Audio, Image and Video Analysis: Theory and Applications* (Vol.1, pp. 1–22). Springer.
14. Caron, M., Bojanowski, P., Joulin, A., & Douze, M. (2018). Deep clustering for unsupervised learning of visual features. In *Proceedings of the European Conference on Computer Vision (ECCV)* (Vol.1, pp. 132–149).
15. Caruana, R., & Niculescu-Mizil, A. (2006, June). An empirical comparison of supervised learning algorithms. In *Proceedings of the 23rd International Conference on Machine Learning* (Vol.23, pp. 161–168).
16. Chali, S. R. (2009). Complex Question Answering: Unsupervised Learning Approaches and Experiments. *Journal of Artificial Intelligent Research*, 1–47.
17. Char, D. S., Shah, N. H., & Magnus, D. (2018). Implementing machine learning in health care—addressing ethical challenges. *The New England journal of medicine, 378*(11), 981.
18. Chen, Y., Wang, J. Z., & Krovetz, R. (2005). CLUE: cluster-based retrieval of images by unsupervised learning. *IEEE Transactions on Image Processing, 14*(8), 1187–1201.
19. Chetlur, S., Woolley, C., Vandermersch, P., Cohen, J., Tran, J., Catanzaro, B., & Shelhamer, E. (2014). cudnn: Efficient primitives for deep learning (Vol.1, pp. 1–22). *arXiv preprint arXiv:1410.0759.*
20. Cho, S., Asfour, S., Onar, A., & Kaundinya, N. (2005). Tool breakage detection using support vector machine learning in a milling process. *International Journal of Machine Tools and Manufacture, 45*(3), 241–249.
21. Cios, K. J., Swiniarski, R. W., Pedrycz, W., & Kurgan, L. A. (2007). Unsupervised learning: association rules. In *Data Mining* (Vol.1, pp. 289–306). Springer, Boston, MA.

22. Claveau, V., & L'Homme, M. C. (2005). Structuring terminology using analogy-based machine learning. In *Proceedings of the 7th International Conference on Terminology and Knowledge Engineering, TKE* (Vol.1, pp. 17–18).
23. Criminisi, A., Shotton, J., & Konukoglu, E. (2012). Decision forests: A unified framework for classification, regression, density estimation, manifold learning and semi-supervised learning. *Foundations and Trends® in Computer Graphics and Vision*, *7*(2–3), 81–227.
24. DasGupta, A. (2011). *Probability for Statistics and Machine Learning: Fundamentals and Advanced Topics*. Springer Science & Business Media (Vol.1, pp. 1–12).
25. Dasgupta, A., Sun, Y. V., König, I. R., Bailey-Wilson, J. E., & Malley, J. D. (2011). Brief review of regression-based and machine learning methods in genetic epidemiology: the Genetic Analysis Workshop 17 experience. *Genetic Epidemiology*, *35*(S1), S5-S11.
26. Dempster, A. P., Laird, N. M., & Rubin, D. B. (1977). Maximum probability from incomplete data via the EM algorithm. *Journal of the Royal Statistical Society. Series B (Methodological)*, 1–38.
27. Deng, L., & Yu, D. (2014). Deep learning: methods and applications. *Foundations and Trends in Signal Processing*, *7*(3–4), 197–387.
28. Denoeux, T. (2008). A k-nearest neighbor classification rule based on Dempster-Shafer theory. In *Classic Works of the Dempster-Shafer Theory of Belief Functions* (Vol.1, pp. 737–760). Springer, Berlin, Heidelberg.
29. Dietterich, T. G. (2000, June). Ensemble methods in machine learning. In *International Workshop on Multiple Classifier Systems* (Vol.1, pp. 1–15). Springer, Berlin, Heidelberg.
30. Doan, A., Madhavan, J., Domingos, P., & Halevy, A. (2004). Ontology matching: A machine learning approach. In *Handbook on Ontologies* (Vol.1, pp. 385–403). Springer, Berlin, Heidelberg.
31. Du, K. L., & Swamy, M. N. S. (2014). Fundamentals of machine learning. In *Neural Networks and Statistical Learning* (Vol.1, pp. 15–65). Springer, London.
32. Dudani, S. A. (1976). The distance-weighted k-nearest-neighbor rule. *IEEE Transactions on Systems, Man, and Cybernetics*, (4), 325–327.

33. Dutra da Silva, R., Robson, W., & Pedrini Schwartz, H. (2011). Image segmentation based on wavelet feature descriptor and dimensionality reduction applied to remote sensing. *Chilean J. Stat, 2*.
34. Fausett, L. (19994). *Fundamentals of Neural Networks.* New York: Prentice Hall.
35. Figueiredo, M. A. T., & Jain, A. K. (2002). Unsupervised learning of finite mixture models. *IEEE Transactions on Pattern Analysis and Machine Intelligence*, *24*(3), 381–396.
36. Forsyth, R. S. (1990). The strange story of the Perceptron. *Artificial Intelligence Review*, *4*(2), 147–155.
37. Franc, V., & Hlaváč, V. (2005, August). Simple solvers for large quadratic programming tasks. In *Joint Pattern Recognition Symposium* (pp. 75–84). Springer, Berlin, Heidelberg.
38. Freund, Y., & Mason, L. (1999, June). The alternating decision tree learning algorithm. In *ICML* (Vol. 99, pp. 124–133).
39. Freund, Y., & Schapire, R. E. (1999). Large margin classification using the perceptron algorithm. *Machine Learning*, *37*(3), 277–296.
40. Fukunaga, A. S. (2008). Automated discovery of local search heuristics for satisfiability testing. *Evolutionary Computation*, *16*(1), 31–61.
41. Gaddam, S. R., Phoha, V. V., & Balagani, K. S. (2007). K-Means+ ID3: A novel method for supervised anomaly detection by cascading K-Means clustering and ID3 decision tree learning methods. *IEEE Transactions on Knowledge and Data Engineering*, *19*(3), 345–354.
42. Garcia, V., Debreuve, E., & Barlaud, M. (2008, June). Fast k nearest neighbor search using GPU. In *2008 IEEE Computer Society Conference on Computer Vision and Pattern Recognition Workshops* (Vol.1, pp. 1–6). IEEE.
43. Ge, Z., Song, Z., Ding, S. X., & Huang, B. (2017). Data mining and analytics in the process industry: The role of machine learning. *Ieee Access*, *5*, 20590–20616.
44. González, L., Angulo, C., Velasco, F., & Catala, A. (2006). Dual unification of bi-class support vector machine formulations. *Pattern Recognition*, *39*(7), 1325–1332.
45. Goswami, R., Dufort, P., Tartaglia, M. C., Green, R. E., Crawley, A., Tator, C. H., ... & Davis, K. D. (2016). Frontotemporal correlates of impulsivity and machine learning in retired professional athletes with a

history of multiple concussions. *Brain Structure and Function*, *221*(4), 1911–1925.

46. Gross, G. N., Lømo, T., & Sveen, O. (1969). Participation of inhibitory and excitatory interneurones in the control of hippocampal cortical output, Per Anderson, The Interneuron.
47. Grzymala-Busse, J. W., & Hu, M. (2000). A comparison of several approaches to missing attribute values in data mining. In *International Conference on Rough Sets and Current Trends in Computing* (pp. 378–385). Springer, Berlin, Heidelberg.
48. Grzymala-Busse, J. W., Goodwin, L. K., Grzymala-Busse, W. J., & Zheng, X. (2005). Handling missing attribute values in preterm birth data sets. In *International Workshop on Rough Sets, Fuzzy Sets, Data Mining, and Granular-Soft Computing* (pp. 342–351). Springer, Berlin, Heidelberg.
49. Guzella, T. S., & Caminhas, W. M. (2009). A review of machine learning approaches to spam filtering. *Expert Systems with Applications*, *36*(7), 10206–10222.
50. Haykin, S. (1994). *Neural Networks: A Comprehensive Foundation.* New York: Macmillan Publishing (Vol.1, pp. 30-59).
51. Haykin, S., & Network, N. (2004). A comprehensive foundation. *Neural Networks*, *2*(2004), 41.
52. Herbrich, R. (2017, February). Machine Learning at Amazon. In *WSDM* (Vol.1, pp. 520–535).
53. Hofmann, T. (2001). Unsupervised learning by probabilistic latent semantic analysis. *Machine Learning*, *42*(1–2), 177–196.
54. Holmes, G., Donkin, A., & Witten, I. H. (1994, November). Weka: A machine learning workbench. In *Proceedings of ANZIIS'94-Australian New Zealnd Intelligent Information Systems Conference* (Vol.1, pp. 357–361). IEEE.
55. Honghai, F., Guoshun, C., Cheng, Y., Bingru, Y., & Yumei, C. (2005, September). A SVM regression based approach to filling in missing values. In *International Conference on Knowledge-Based and Intelligent Information and Engineering Systems* (pp. 581–587). Springer, Berlin, Heidelberg.
56. Hornik, K., Stinchcombe, M., & White, H. (1989). Multilayer feedforward networks are universal approximators. *Neural Networks*, *2*(5), 359–366.

57. Jain, A. K., Mao, J., & Mohiuddin, K. M. (1996). Artificial neural networks: A tutorial. *Computer*, *29*(3), 31-44.
58. Jain, P., Garibaldi, J. M., & Hirst, J. D. (2009). Supervised machine learning algorithms for protein structure classification. *Computational Biology and Chemistry*, *33*(3), 216–223.
59. Jordan, M. I., & Mitchell, T. M. (2015). Machine learning: Trends, perspectives, and prospects. *Science*, *349*(6245), 255–260.
60. Kaelbling, L. P., Littman, M. L., & Moore, A. W. (1996). Reinforcement learning: A survey. *Journal of Artificial Intelligence Research*, *4*, 237–285.
61. Kamiran, F., Calders, T., & Pechenizkiy, M. (2010, December). Discrimination aware decision tree learning. In *2010 IEEE International Conference on Data Mining* (Vol.1, pp. 869–874). IEEE.
62. Kearns, M., & Mansour, Y. (1999). On the boosting ability of top–down decision tree learning algorithms. *Journal of Computer and System Sciences*, *58*(1), 109–128.
63. Kelleher, J. D. (2016, October). Fundamentals of machine learning for neural machine translation. In *Proceedings of the Translating Europen Forum* (Vol.1, pp. 1–42).
64. Kelleher, J. D., Mac Namee, B., & D'arcy, A. (2015). *Fundamentals of Machine Learning for Predictive Data Analytics: Algorithms, Worked Examples, and Case Studies* (Vol.1, pp. 1–32). MIT Press.
65. Keller, J. M., Gray, M. R., & Givens, J. A. (1985). A fuzzy k-nearest neighbor algorithm. *IEEE Transactions on Systems, Man, and Cybernetics*, (4), 580–585.
66. Keller, R. E., & Poli, R. (2007, September). Linear genetic programming of parsimonious metaheuristics. In *Evolutionary Computation, 2007. CEC 2007. IEEE Congress on* (pp. 4508–4515). IEEE.
67. Kenett, R. S., & Shmueli, G. (2015). Clarifying the terminology that describes scientific reproducibility. *Nature Methods*, *12*(8), 699–699.
68. King, D. E. (2009). Dlib-ml: A machine learning toolkit. *The Journal of Machine Learning Research*, *10*, 1755–1758.
69. Kingma, D. P., Mohamed, S., Rezende, D. J., & Welling, M. (2014). Semi-supervised learning with deep generative models. In *Advances in Neural Information Processing Systems* (Vol.1, pp. 3581–3589).
70. Kohonen, T. (1997). Self-Organizating Maps.

71. Kononenko, I. (2001). Machine learning for medical diagnosis: history, state of the art and perspective. *Artificial Intelligence in Medicine, 23*(1), 89–109.
72. Kotsiantis, S. B., Zaharakis, I., & Pintelas, P. (2007). Supervised machine learning: A review of classification techniques. *Emerging Artificial Intelligence Applications in Computer Engineering, 160*(1), 3–24.
73. Kumar, V., Kalitin, D., & Tiwari, P. (2017, May). Unsupervised learning dimensionality reduction algorithm PCA for face recognition. In *2017 International Conference on Computing, Communication, and Automation (ICCCA)* (Vol.1, pp. 32–37). IEEE.
74. Kumar, V., Verma, A., Mittal, N., & Gromov, S. V. (2019). Anatomy of preprocessing of big data for monolingual corpora paraphrase extraction: source language sentence selection. In *Emerging Technologies in Data Mining and Information Security* (Vol.1, pp. 495–505). Springer, Singapore.
75. Lavecchia, A. (2015). Machine-learning approaches in drug discovery: methods and applications. *Drug discovery today, 20*(3), 318–331.
76. LeCun, Y., Bengio, Y., & Hinton, G. (2015). Deep learning. *Nature, 521*(7553), 436–444.
77. Li, D., Deogun, J., Spaulding, W., & Shuart, B. (2004, June). Towards missing data imputation: a study of fuzzy k-means clustering method. In *International Conference on Rough Sets and Current Trends in Computing* (pp. 573–579). Springer, Berlin, Heidelberg.
78. Li, M., Zhu, X., & Gong, S. (2018). Unsupervised person re-identification by deep learning tracklet association. In *Proceedings of the European Conference on Computer Vision (ECCV)* (Vol.1, pp. 737–753).
79. Liu, Y., Zhang, D., & Lu, G. (2008). Region-based image retrieval with high-level semantics using decision tree learning. *Pattern Recognition, 41*(8), 2554–2570.
80. Luengo, J., García, S., & Herrera, F. (2012). On the choice of the best imputation methods for missing values considering three groups of classification methods. *Knowledge and Information Systems, 32*(1), 77–108.
81. Luis Gonz, l. A. (2005). Unified dual for bi-class SVM approaches. *Pattern Recognition, 38* (10), 1772–1774.

82. Michie, D., Spiegelhalter, D. J., & Taylor, C. C. (1994). Machine learning. *Neural and Statistical Classification*, *13*(1994), 1–298.

83. Mnih, V., Kavukcuoglu, K., Silver, D., Rusu, A. A., Veness, J., Bellemare, M. G., ... & Petersen, S. (2015). Human-level control through deep reinforcement learning. *Nature*, *518*(7540), 529–533.

84. Monostori, L., Márkus, A., Van Brussel, H., & Westkämpfer, E. (1996). Machine learning approaches to manufacturing. *CIRP Annals*, *45*(2), 675–712.

85. Mooney, S. J., & Pejaver, V. (2018). Big data in public health: terminology, machine learning, and privacy. *Annual Review of Public Health*, *39*, 95–112.

86. Muggleton, S., & De Raedt, L. (1994). Inductive logic programming: Theory and methods. *The Journal of Logic Programming*, *19*, 629–679.

87. Navigli, R., Velardi, P., & Gangemi, A. (2003). Ontology learning and its application to automated terminology translation. *IEEE Intelligent Systems*, *18*(1), 22–31.

88. Ng, V., & Cardie, C. (2002, July). Improving machine learning approaches to coreference resolution. In *Proceedings of the 40th annual meeting of the Association for Computational Linguistics* (pp. 104–111).

89. Nielsen, H., Brunak, S., & von Heijne, G. (1999). Machine learning approaches for the prediction of signal peptides and other protein sorting signals. *Protein Engineering*, *12*(1), 3–9.

90. Novikoff, A. B. (1963). *On Convergence Proofs for Perceptrons*. Stanford Research Inst Menlo Park, CA.

91. Olden, J. D., Lawler, J. J., & Poff, N. L. (2008). Machine learning methods without tears: a primer for ecologists. *The Quarterly review of biology*, *83*(2), 171–193.

92. Olga, R., Deng, J., Su, H., Krause, J., Satheesh, S., Ma, S., & Berg, A. C. (2015). Large scale visual recognition challenge. *ImageNet http:// arxiv. org/abs/1409.0575*.

93. Oltean, M. (2007). Evolving evolutionary algorithms with patterns. *Soft Computing*, *11*(6), 503–518.

94. Oltean, M., & Dumitrescu, D. (2004). Evolving TSP heuristics using multi expression programming. In *International Conference on Computational Science* (pp. 670–673). Springer, Berlin, Heidelberg.

95. Oltean, M., & Groşan, C. (2003, September). Evolving evolutionary algorithms using multi expression programming. In *European Conference on Artificial Life* (pp. 651–658). Springer, Berlin, Heidelberg.
96. Parmar, C., Grossmann, P., Bussink, J., Lambin, P., & Aerts, H. J. (2015). Machine learning methods for quantitative radiomic biomarkers. *Scientific Reports*, *5*, 13087.
97. Pedregosa, F., Varoquaux, G., Gramfort, A., Michel, V., Thirion, B., Grisel, O., ... & Vanderplas, J. (2011). Scikit-learn: Machine learning in Python. *The Journal of Machine Learning Research*, *12*, 2825–2830.
98. Poli, R., Woodward, J., & Burke, E. K. (2007, September). A histogram-matching approach to the evolution of bin-packing strategies. In *Evolutionary Computation, 2007. CEC 2007. IEEE Congress on* (pp. 3500–3507). IEEE.
99. Price, J., Buckles, K., Van Leeuwen, J., & Riley, I. (2019). *Combining Family History and Machine Learning to Link Historical Records* (No. w26227, pp. 1–33). National Bureau of Economic Research.
100. Rasmussen, C. E. (2003, February). Gaussian processes in machine learning. In *Summer School on Machine Learning* (Vol.1, pp. 63–71). Springer, Berlin, Heidelberg.
101. Rosenblatt, F. (1961). Principles of Neurodynamics Unclassifie— Armed Services Technical Informatm Agency. Spartan, Washington, DC.
102. Rouet-Leduc, B., Hulbert, C., Lubbers, N., Barros, K., Humphreys, C. J., & Johnson, P. A. (2017). Machine learning predicts laboratory earthquakes. *Geophysical Research Letters*, *44*(18), 9276–9282.
103. Sammut, C., & Webb, G. I. (2017). *Encyclopedia of Machine Learning and Data Mining* (Vol.1, pp. 1–34). Springer.
104. Schaffalitzky, F., & Zisserman, A (2004). Automated scene matching in movies. CIVR 2004. *Proceedings of the Challenge of Image and Video Retrieval, London, LNCS, 2383.*
105. Schlegl, T., Seeböck, P., Waldstein, S. M., Schmidt-Erfurth, U., & Langs, G. (2017, June). Unsupervised anomaly detection with generative adversarial networks to guide marker discovery. In *International Conference on Information Processing in Medical Imaging* (Vol.1, pp. 146–157). Springer, Cham.

106. Schmid, S. R., Hamrock, B. J., & Jacobson, B. O. (2014). *Fundamentals of Machine Elements: SI Version*. CRC Press (Vol.1, pp. 1–22).

107. Schmidhuber, J. (2015). Deep learning in neural networks: An overview. *Neural Networks*, *61*, 85–117.

108. Sclano, F., & Velardi, P. (2007). Termextractor: a web application to learn the shared terminology of emergent web communities. In *Enterprise Interoperability II* (Vol.2, pp. 287–290). Springer, London.

109. Sebastiani, F. (2002). Machine learning in automated text categorization. *ACM Computing Surveys (CSUR)*, *34*(1), 1–47.

110. Shami, M., & Verhelst, W. (2007). An evaluation of the robustness of existing supervised machine learning approaches to the classification of emotions in speech. *Speech Communication*, *49*(3), 201–212.

111. Shrestha, D. L., & Solomatine, D. P. (2006). Machine learning approaches for estimation of prediction interval for the model output. *Neural Networks*, *19*(2), 225–235.

112. Snelson, E. L. (2007). *Flexible and Efficient Gaussian Process Models for Machine Learning*, Doctoral dissertation (Vol.1, pp. 1–22), UCL (University College London).

113. Song, S. K., Choi, Y. S., Chun, H. W., Jeong, C. H., Choi, S. P., & Sung, W. K. (2011, December). Multi-words terminology recognition using web search. In *International Conference on U-and E-Service, Science and Technology* (Vol.1, pp. 233–238). Springer, Berlin, Heidelberg.

114. Srinivasan, K., & Fisher, D. (1995). Machine learning approaches to estimating software development effort. *IEEE Transactions on Software Engineering*, *21*(2), 126–137.

115. Tan, S. (2005). Neighbor-weighted k-nearest neighbor for unbalanced text corpus. *Expert Systems with Applications*, *28*(4), 667–671.

116. Tavares, J., Machado, P., Cardoso, A., Pereira, F. B., & Costa, E. (2004, April). On the evolution of evolutionary algorithms. In *European Conference on Genetic Programming* (pp. 389–398). Springer, Berlin, Heidelberg.

117. Teather, L. A. (2006). Pathophysiological effects of inflammatory mediators and stress on distinct memory systems. In *Nutrients, Stress, and Medical Disorders* (pp. 377–386). Humana Press.

118. Timothy Jason Shepard, P. J. (1998). Decision Fusion Using a Multi-Linear Classifier. In *Proceedings of the International Conference on Multisource-Multisensor Information Fusion*.

119. Vartak, M., Subramanyam, H., Lee, W. E., Viswanathan, S., Husnoo, S., Madden, S., & Zaharia, M. (2016, June). ModelDB: a system for machine learning model management. In *Proceedings of the Workshop on Human-In-the-Loop Data Analytics* (Vol.1, pp. 1–3).
120. Vellido, A., Martín-Guerrero, J. D., & Lisboa, P. J. (2012, April). Making machine learning models interpretable. In *ESANN* (Vol. 12, pp. 163–172).
121. Voyant, C., Notton, G., Kalogirou, S., Nivet, M. L., Paoli, C., Motte, F., & Fouilloy, A. (2017). Machine learning methods for solar radiation forecasting: A review. *Renewable Energy*, *105*, 569–582.
122. Wilson, A. C., Roelofs, R., Stern, M., Srebro, N., & Recht, B. (2017). The marginal value of adaptive gradient methods in machine learning. In *Advances in Neural Information Processing Systems* (Vol. 1, pp. 4148–4158).
123. Witschel, H. F. (2005). Terminology extraction and automatic indexing. *Terminology and Content Development*, 1, 350–363.
124. Ye, Q., Zhang, Z., & Law, R. (2009). Sentiment classification of online reviews to travel destinations by supervised machine learning approaches. *Expert Systems with Applications*, *36*(3), 6527–6535.
125. Yoganarasimhan, H. (2020). Search personalization using machine learning. *Management Science*, *66*(3), 1045–1070.
126. Yu, X., Pu, K. Q., & Koudas, N. (2005, April). Monitoring k-nearest neighbor queries over moving objects. In *21st International Conference on Data Engineering (ICDE'05)* (Vol.1, pp. 631–642). IEEE.
127. Yusupov, T. (2007). *The Efficient Market Hypothesis Through the Eyes of An Artificial Technical Analyst* (Doctoral dissertation, Christian-Albrechts Universität Kiel).
128. Zhang, H., Goodfellow, I., Metaxas, D., & Odena, A. (2019, May). Self-attention generative adversarial networks. In *International Conference on Machine Learning* (Vol.1, pp. 7354–7363).
129. Zhang, L., Tan, J., Han, D., & Zhu, H. (2017). From machine learning to deep learning: progress in machine intelligence for rational drug discovery. *Drug Discovery Today*, *22*(11), 1680–1685.
130. Zhu, X. J. (2005). *Semi-Supervised Learning Literature Survey*. University of Wisconsin-Madison Department of Computer Sciences.

131. Zhu, X., & Goldberg, A. B. (2009). Introduction to semi-supervised learning. *Synthesis Lectures on Artificial Intelligence and Machine Learning*, *3*(1), 1–130.
132. Zhu, X., Ghahramani, Z., & Lafferty, J. D. (2003). Semi-supervised learning using gaussian fields and harmonic functions. In *Proceedings of the 20th International Conference on Machine Learning (ICML-03)* (pp. 912–919).
133. Zoltan-Csaba, M., Pangercic, D., Blodow, N., & Beetz, M. (2011). Combined 2D-3D categorization and classification for multimodal perception systems. *Int. J. Robotics Res. Arch*, *30*(11).
134. Zuccala, A., van Someren, M., & van Bellen, M. (2014). A machine-learning approach to coding book reviews as quality indicators: Toward a theory of megacitation. *Journal of the Association for Information Science and Technology*, *65*(11), 2248–2260.

Chapter 2

Physical Aspects of Machine Learning in Data Science

CONTENTS

2.1. INTRODUCTION

In numerous industries, data science is now familiar. Some core elements of this ubiquity are the amalgamation of computing in general, the advancement in machine learning with the escalating accessibility of data (Varshney & Alemzadeh, 2017; Dartmann et al., 2019).

Sectors such as marketing and e-commerce have activated machine learning, which has long been sought out for business objectives. Algorithms are employed to spur scientific inventions in natural sciences and we see increasing applications in it. Machine learning uses in natural sciences have been noted in a brief, non-comprehensive list below (more details can be found in Karpatne et al. (2017) and Carleo et al. (2019):

- *Biology*: classification of gene-to-protein, DNA/RNA matching, neural designs.
- *Materials*: fresh compounds being found, material characterization (formation -> property linkage).
- *Fluid mechanics*: accuracy of numerical simulations, particularly in crucial regimes like turbulence, etc. are maximized.
- *Quantum mechanics*: DFT approximation.
- *Energy*: enhancement of yield, formation, and consumption forecast.
- *Climate & geology*: better coupling amongst subprocesses, multi-scale modeling on time and space.
- *Particle physics*: finding of the Higg's boson is the best success, rare occasions detection.

Nevertheless, throwing the current machine learning algorithm together is not enough. Physical occurrences for which we only have flawed models in the first place are being examined here. Those mathematical designs can be bettered with data science. We must make sure that the machine learning productions are physically stable to be of practical use. To combine knowledge into machine learning, or in other words, when we marry information and natural sciences together, special attention must be given (VanderPlas et al., 2012; Way et al., 2012).

Physicists and engineers from distinct fields have created the literature that has been collected by us. Some have examined the results of the others to take knowledge from those many experiments, while other authors have performed actual experiments, compared *data-based* with *physics-based* approaches. Factors that must lead us when putting data science to satisfying use according to them are here (Swan, 2013; Holzinger & Jurisica, 2014):

- *Efficiency*—when in comparison with actual data, the algorithm must make good forecasts (this is not particular to natural sciences obviously).
- *Generalizability*—Other than just the analysis under study, the algorithm must have caught "something" of a more standard phenomenon.
- *Interpretability*—To give a better comprehension of the physical occurrence at play, that "something" can be derived from the algorithmic model (interpretability is a severe presumption).
- *Common language, tools, and platforms*—physicists from various fields must be able to work together by exchanging insights, information, and experiments hence, the algorithm must be incorporated into a global structure (this is critical for instance in multi-scale setups). The association must reach other people such as engineers, who will be behind the industrialization if we go after developing a fresh product (Krefl & Seong, 2017; Bergen et al., 2019).

In their area of expertise, engineers and physicists are having a crucial view on the employment of data science. They recognize their complementarity and do not deny the empirical (data-based) and inductive (physics-based) schemes *(In a precise context of the current study, for a large number of figures from measurements and simulations, the machine learning is described as the ability to make efficient surrogates* (Chang & Dinh, 2019)).

2.2. THEORY-GUIDED DATA SCIENCE

The fusion of physics and data science can be done in several ways. The will to capitalize on present practices has given birth to the concept of Theory-Guided Data Science which is thoroughly described in the Karpatne et al. (2017) paper. Amongst the two fields, TGDS can be seen as a cry for mutual awareness. The necessity to create a catalog of pleasant practices for accomplishing the stakes cited at the beginning of this article: *"The overarching vision of TGDS is to introduce scientific consistency as an essential component for learning generalizable models,"* and the growing significance of data in natural science is admitted by it.

In Karpatne et al. (2017), the performance of a machine learning model is defined as:

Performance ∝ Accuracy + Simplicity + Consistency

- *Accuracy*, when doing machine learning is what we're generally after- to precisely reflect the division of the data we want to foretell should be the potential of the model (Niebert et al., 2012; Zheng et al., 2019).
- *Simplicity* in data science, associates with the bias/variance trade-off regularly put forward. The "best" designs are those that stabilize precision with robustness against the inconsistency of the training information, rather than the most complicated ones (Buhi & Goodson, 2007; Sheikh & Jahirabadkar, 2018).
- The element that embeds all the strength of TGDS is *consistency*. The machine learning models that were not responsible to be consistent with the basic physical theory and thus, generalize weakly is what it protects against (the declaration that consistency can decrease variance without impairing preference should be taken lightly given the informal depiction of the concepts that the paper states).

The driving notion behind the many practices described in Karpatne et al. (2017) is consistency. (Faghmous & Kumar, 2014; Wagner & Rondinelli, 2016) are the five TGDS research concepts the practices are arranged in:

- Augmenting theory-based models employing data science.
- Learning hybrid models of theory and data science.
- Theory-guided refinement of data science production.
- Theory-guided training of data science models.
- Theory-guided design of data science models.

2.2.1. Theory-Guided Design of Data Science Models

The architecture, (when workable) the link task among the inputs and the product, and the selection of the machine learning algorithm are guided by the domain information in this theme (Schmidt et al., 1995; Shi & Levy, 2005; Karpatne et al., 2017).

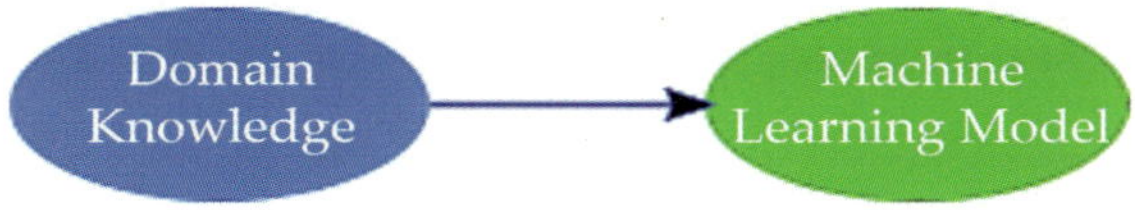

Figure 2.1. Theory guided data science design's input-output model.

[Source: https://towardsdatascience.com/weaving-machine-learning-into-physics-a-data-scientists-point-of-view-63f60a95f5be.]

An instance is that, because the issue under scrutiny consists of spatial correlations, the scientist may be led towards a fake neural network, with convolutional membranes (Sullivan et al., 2007; Friák et al., 2015; Zheng et al., 2019).

2.2.2. Theory-Guided Training of Data Science Models

The data science model has many constraints enforced upon it which are guided by domain knowledge (symmetries, bounds for values or gradients, regularization, the interaction between variables, preliminary values of parameters). "Model selection" is what it is placed with (Kliegl & Baltes, 1987; Nadeem et al., 2016).

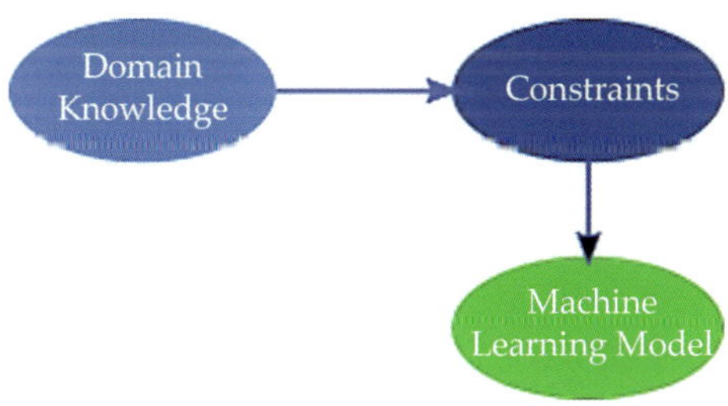

Figure 2.2. Theory-guided training model theory-guided.

[Source: https://towardsdatascience.com/weaving-machine-learning-into-physics-a-data-scientists-point-of-view-63f60a95f5be.]

For instance, a penalization period can be added to the loss role of the machine learning model respectively if the forecast target must take its figures in a known range. The surplus training test piece can be unnaturally made to coerce the model to display the same invariance if the issue is invariant under some change of the input space (Brandmaier et al., 2016; Sun et al., 2020).

2.2.3. Theory-Guided Refinement of Data Science Outputs

To comprehend them or to make them physically stable, the product of the machine learning model is post-processed (polished) according to theoretical considerations (Reichstein et al., 2019).

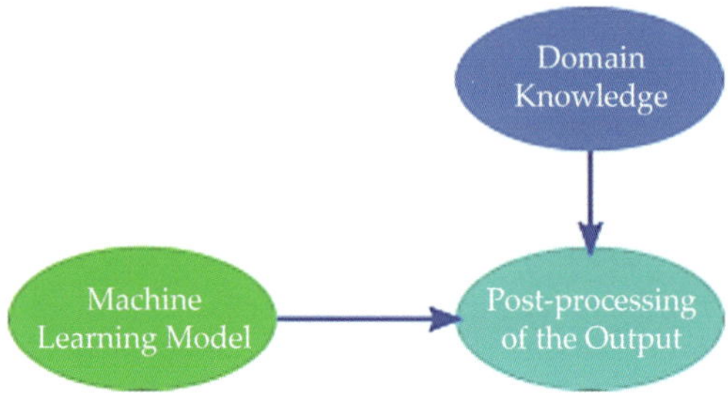

Figure 2.3. Theory-guided refinement model.

[Source: https://towardsdatascience.com/weaving-machine-learning-into-physics-a-data-scientists-point-of-view-63f60a95f5be.]

To easily remove a candidate model that outputs inconsistent results is the easiest example of it. With the faith that it meets the consistency conditions, another algorithm or group of parameters can then be investigated. The outcomes may be "tweaked," like by interchanging aberrant figures with something more rational if the inconsistencies are not too serious (Faber et al., 2017).

2.2.4. Learning Hybrid Models of Theory and Data Science

With machine learning models, theoretical and/or numerical models coexist. Conversely, the products of machine learning models can incite traditional models, and the latter gives the inputs for the former.

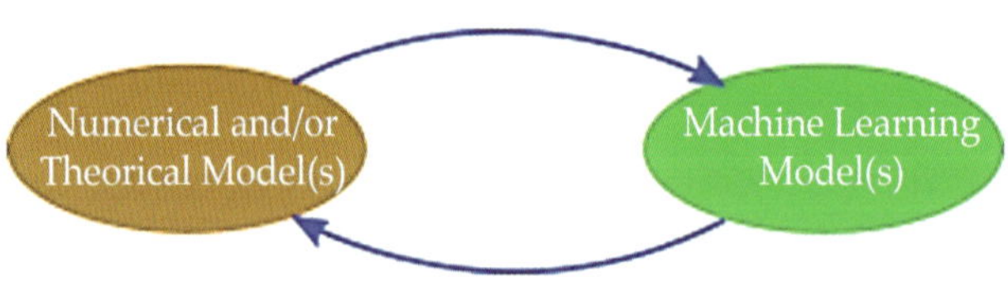

Figure 2.4. Hybrid learning model.

[Source: https://towardsdatascience.com/weaving-machine-learning-into-physics-a-data-scientists-point-of-view-63f60a95f5be.]

Intermediate outputs flow amongst them until the concluding output step and all those models form an intricate graph (Zahedi et al., 2016).

The theory of the *surrogate model* is one test used. Within the system, a theoretical model's unavailability is coped by the surrogate model. The model learns the link among the variables involved instead of taking equations out of thin air and is connected to a more standard theoretical or numerical model. When the theoretical model is within reach but too expensive to run numerically on every occasion, a surrogate model proves useful then (Squire & Jan 2007).

2.2.5. Augmenting Theory-Based Models Utilizing Data Science

Machine learning models are not essentially built here. To assist the theoretician better their models, data science practices are called in (Blei & Smyth, 2017).

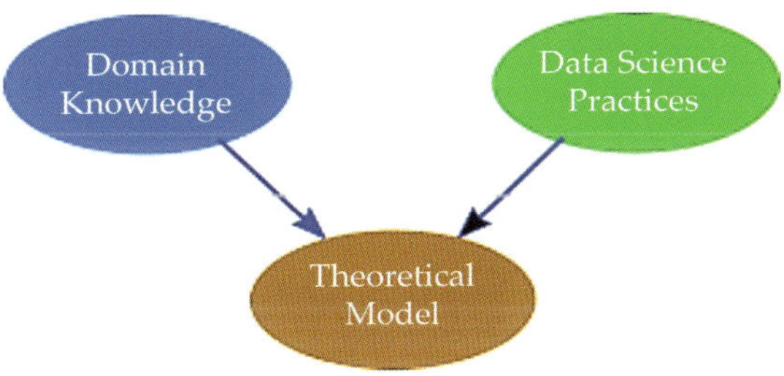

[Source: https://towardsdatascience.com/weaving-machine-learning-into-physics-a-data-scientists-point-of-view-63f60a95f5be.]

Figure 2.5. Augmenting theory-based model.

Data assimilation and parameter calibration are the two use cases explained in Karpatne et al. (2017) for this theme. To increase the possibility of the theoretical model or its parameters, they depend on inference and uncertainty modeling (Finzer, 2013).

2.3. WEAVING MOTIFS

The work is not commonly planned this way by data scientists. As mentioned below for supervised learning, a data science workflow or pipeline is employed (Van Der Aalst, 2016; Samulowitz et al., 2018):

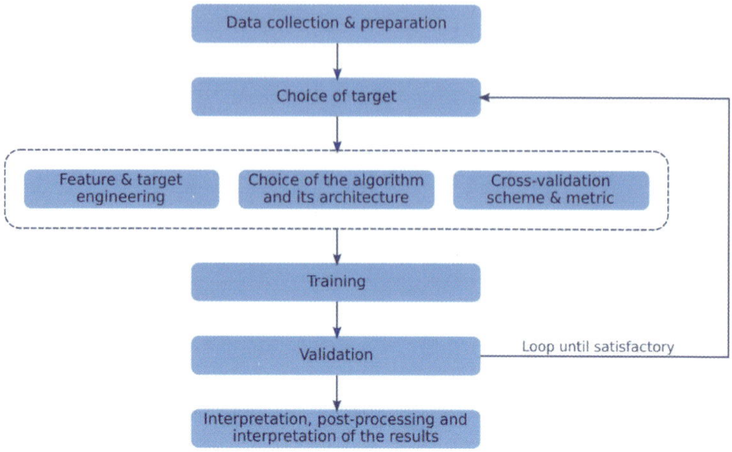

Figure 2.6. Data science workflow for supervised learning.

[Source: https://towardsdatascience.com/weaving-machine-learning-into-physics-a-data-scientists-point-of-view-63f60a95f5be.]

As ways to present domain knowledge at distinct points of our data science workflow, let's examine the TGDS practices. Refer to the Karpatne et al. (2017) paper for instances of practical knowledge that can be collected in such situations.

- According to Karpatne et al. (2017), "multi-task learning" is the enhanced form of the partitioning of input details. For every partition, it allows a distinct machine learning design. In the table, the practice is divided into two rows. The partitioning criterion is exactly dictated by domain knowledge in a data preparation stage. The criterion is obtained from the data by exercising an initial model on it in a feature engineering stage (through an unsupervised clustering algorithm or explicit)
- Those practices are not cited in Karpatne et al. (2017) but they sure fit here. For a discussion on "choice of the target" and "cross-validation scheme and metric," which are missing altogether, see the next section. Mine: These ways sure fit here but they are not stated in Karpatne et al. (2017). View the next section for a discussion on "choice of the target" and "cross-validation scheme & metric," which are missing.
- Along with the example of generalized linear models (GLM), this article can be found under the name "specification of response"

in Karpatne et al. (2017). Yet the argument can be stretched to a relationship function given a-priority in algorithms, like support vector machines with a kernel.

- As data assimilation (the employment of observations to correct forecasts at each iteration) is precisely the case for online learning or gradient boosting, we map it to the model/algorithm stage.

Because it is not pivoted on data science, one of the ways of TGDS does not fit in our table. It is the data science results being leveraged in a physical or numerical model (stated in the "learning hybrid models of theory and data science" research theme). The topic of a future article, this type of hybridization will be a specific case of the data/physics coupling approachcs.

To complete the picture, let's mention two techniques that are detailed in Ling et al. (2016), inspired by physical considerations. Those techniques aim at producing machine learning models that are more accurate yet simpler (and faster) than "naive" ones. The first technique is to train the machine learning model, not on raw data, but input features inspired by the invariants of the mathematical model. The second one is well known in image recognition applications; it consists of augmenting the input data by applying transformations under which the mathematical model is invariant (e.g., by translating or rotating the inputs).

Two practiccs that arc cxplaincd in Ling ct al. (2016), inspircd by physical considerations will be stated to finish the picture. The production of machine learning designs that are more exact yet simpler (and faster) than "naive" ones is the purpose of this tcchniquc. Thc initial tcchniquc is to teach the machine learning design on input characteristics inspired by the invariants of the mathematical design rather than on raw figures. The initial one is renowned in image recognition application; augmenting the input information by applying changes under which the mathematical design is invariant (e.g., by translating or rotating the inputs) consists of it.

2.4. SIGNIFICANCE OF TGDS

It was rather simple to relay the TGDS practices onto the data science workflow. Amongst data and natural scientists, it is a good prompt in favor of the wanted collaboration.

The fact that some practices were far from naïve was discovered: data science responsibilities such as algorithm tuning and feature engineering are considered ordinary and they go well beyond it. At the training phase, "Regularization and penalization terms (B)" is, arguably, the most informative item. Instead of the method being used to discriminate posterior among probable models, it is one whereby physical consistency is *embedded* inside the machine learning design. Concretely, it incorporates several tricks (Zimmerman, 2008):

- On the factors or their derivatives (gradient), upper and lower limits are put.
- To compel interaction among variables that are known to interact, group lasso regularization is utilized (to put it another way, this is a method of inserting physical causality in the model).
- The deficiency of consistency in some penalizing models—for instance, the water surface must be beneath that of nearby land in modeling landscape elevation.

This practice has also an effect on the "Choice of algorithm and its architecture" stage of the workflow as not all training algorithms or structures allow the data scientist to perform freely on the loss function. For instance, XGBoost lets you connect your objective function and metric. In contrast, a group lasso implementation is not found in the Scikit-learn Python ecosystem.

On the negative side, the choice of the target, and the choice of the cross-validation scheme and metric are two vital (supervised) machine learning tasks that are not stated in Karpatne et al. (2017).

In a given scientific setting, one might ponder the choice of the target as clear: a discrete parameter like a regime, a noticeable quantity, an anticipated outcome such as a yield... are all normal ones. We must select one out of numerous quantities that are connected, several of which could be observable in the existence of a mathematical model. If we contemplate that all candidate targets may not display the same measurement incertitude or data quality problems, our mathematical model is flawed and that the observations can be noisy, then they are not of like interest.

We would imagine this item to be addressed plainly as regards cross-validation in a data science context. We can deduct that the question was considered too apparent by the authors of Karpatne et al. (2017) like the choice of the target above. The RMSE (Root-Mean Squared Error) is by far the most pleasant choice for regressions for actual physical problems. The precision feels just as natural for one-class classification complications. The part of cross-validation in the capability of the model to generalize (and thus in its consistency) must be recognized. Samples of knowledge that the cross-validation structure and metrics are susceptible to encode are given here:

- Temporal and/or spatial type of the phenomena (dodging leaks by correctly dividing the input data set).
- The target being statistically disseminated.
- Extreme factors being handled or “squashed” (such as come across noisy or unfinished observations).
- Inconsistent models being penalized.

The definition of the target and cross-validation are two matters not to be ignored. As the two ideas have both physical and methodological suggestions, a data scientist and a physicist working together would need to be on par to debate them. Here, this is not just the physicist asking the data scientist to provide their knowledge.

2.5. SHARING AND REUSING KNOWLEDGE WITH DATA SCIENCE

Approaches for fusing data science with physics has been the emphasis of our discussion so far. Efficiency, Generalizability, Interpretability are the primary three ones among the stakes enumerated in the introduction that is content by those approaches. A call for an interdisciplinary association is what the last stake—Common language, tools & platforms—has been considered as. Let’s be more detailed now (Faniel & Zimmerman, 2011).

Data science is common on many scientific grounds as was already mentioned. It makes a connection between physical phenomena and procedures so it is well recognized in those industries. The entire lifecycle of a product: research, design, process engineering, quality control, supply chain, logistics, customer service, ... has data science as a perfect collaboration vehicle for all events involved in it.

There are at least 3 stages that create or consume artifacts that are appreciated across the organization recalling the data science workflow drawn earlier:

- *Data collection & preparation*: reusable parts of information are data sets, their meaning, the methods used to get them, the outcome of cleaning, their structure, the documentation.
- *Feature & target engineering*: new features are valuable to other teams in the organizations as they are fashioned from existing data which is most probable to have a physical meaning.

- *Interpretation, post-processing, and exploitation of results*: the workflow creates vital artifacts such as data visualizations, interpretative reports, etc. together with new data and a trained model.

Inside a company, the above is valid. As long as they wish to unite, it also holds among organizations working in the same industry to some extent. It is usual nowadays for one to issue one's findings as reusable software packages, for instance, for visualization of the outcomes, etc. in material science, see [PYMKS], it provides roles for calculations on microstructures.

Data is the fuel for building better machine learning models; it is a call for sharing data sets broadly. It can come in many appearances: experiments, measurements, numerical simulations, ... or models themselves. Although it is highly wanted, extensive sharing of data is tough in practice. Apparent obstacles are the absence of an agreement on metadata and intellectual property. A data repository should come with a computation grid, software and a metadata catalog are others that should be recognized. It is indeed an intricate project to create a data platform with many stakeholders. Hill et al. (2016) and Brough et al. (2017) are at least demanding for such a platform.

REFERENCES

1. Bergen, K. J., Chen, T., & Li, Z. (2019). Preface to the focus section on machine learning in seismology. *Seismological Research Letters*, *90*(2A), 477–480.
2. Blei, D. M., & Smyth, P. (2017). Science and data science. *Proceedings of the National Academy of Sciences*, *114*(33), 8689–8692.
3. Brandmaier, A. M., Prindle, J. J., McArdle, J. J., & Lindenberger, U. (2016). Theory-guided exploration with structural equation model forests. *Psychological Methods*, *21*(4), 566.
4. Brough, D. B., Wheeler, D., & Kalidindi, S. R. (2017). Materials knowledge systems in python—a data science framework for accelerated development of hierarchical materials. *Integrating Materials and Manufacturing Innovation*, *6*(1), 36–53.
5. Buhi, E. R., & Goodson, P. (2007). Predictors of adolescent sexual behavior and intention: A theory-guided systematic review. *Journal of Adolescent Health*, *40*(1), 4–21.
6. Carleo, G., Cirac, I., Cranmer, K., Daudet, L., Schuld, M., Tishby, N., ... & Zdeborová, L. (2019). Machine learning and the physical sciences. *Reviews of Modern Physics*, *91*(4), 045002.
7. Chang, C. W., & Dinh, N. T. (2019). Classification of machine learning frameworks for data-driven thermal fluid models. *International Journal of Thermal Sciences*, *135*, 559–579.
8. Dartmann, G., Song, H., & Schmeink, A. (Eds.). (2019). *Big Data Analytics for Cyber Physical Systems: Machine Learning for the Internet of Things, 1, 1–22*. Elsevier.
9. Faber, F. A., Hutchison, L., Huang, B., Gilmer, J., Schoenholz, S. S., Dahl, G. E., ... & Von Lilienfeld, O. A. (2017). Prediction errors of molecular machine learning models lower than hybrid DFT error. *Journal of Chemical Theory and Computation*, *13*(11), 5255–5264.
10. Faghmous, J. H., & Kumar, V. (2014). A big data guide to understanding climate change: The case for theory-guided data science. *Big Data*, *2*(3), 155–163.
11. Faghmous, J. H., Banerjee, A., Shekhar, S., Steinbach, M., Kumar, V., Ganguly, A. R., & Samatova, N. (2014). Theory-guided data science for climate change. *Computer*, *47*(11), 74–78.

12. Faniel, I. M., & Zimmerman, A. (2011). Beyond the data deluge: A research agenda for large-scale data sharing and reuse. *International Journal of Digital Curation*, *6*(1), 58–69.

13. Finzer, W. (2013). The data science education dilemma. *Technology Innovations in Statistics Education*, *7*(2), 1–33.

14. Friák, M., Tytko, D., Holec, D., Choi, P. P., Eisenlohr, P., Raabe, D., & Neugebauer, J. (2015). Synergy of atom-probe structural data and quantum-mechanical calculations in a theory-guided design of extreme-stiffness superlattices containing metastable phases. *New Journal of Physics*, *17*(9), 093004.

15. Hill, J., Mulholland, G., Persson, K., Seshadri, R., Wolverton, C., & Meredig, B. (2016). Materials science with large-scale data and informatics: unlocking new opportunities. *Mrs Bulletin*, *41*(5), 399–409.

16. Holzinger, A., & Jurisica, I. (2014). Knowledge discovery and data mining in biomedical informatics: The future is in integrative, interactive machine learning solutions. In *Interactive Knowledge Discovery and Data Mining in Biomedical Informatics* (Vol.1, pp. 1–18). Springer, Berlin, Heidelberg.

17. Karpatne, A., Atluri, G., Faghmous, J. H., Steinbach, M., Banerjee, A., Ganguly, A., ... & Kumar, V. (2017). Theory-guided data science: A new paradigm for scientific discovery from data. *IEEE Transactions on Knowledge and Data Engineering*, *29*(10), 2318–2331.

18. Karpatne, A., Atluri, G., Faghmous, J. H., Steinbach, M., Banerjee, A., Ganguly, A., ... & Kumar, V. (2017). Theory-guided data science: A new paradigm for scientific discovery from data. *IEEE Transactions on Knowledge and Data Engineering*, *29*(10), 2318–2331.

19. Kliegl, R., & Baltes, P. B. (1987). Theory-guided analysis of mechanisms of development and aging through testing-the-limits and research on expertise, 1, 1–33.

20. Krefl, D., & Seong, R. K. (2017). Machine learning of Calabi-Yau volumes. *Physical Review D*, *96*(6), 066014.

21. Ling, J., Jones, R., & Templeton, J. (2016). Machine learning strategies for systems with invariance properties. *Journal of Computational Physics*, *318*, 22–35.

22. Nadeem, E., Weiss, D., Olin, S. S., Hoagwood, K. E., & Horwitz, S. M. (2016). Using a theory-guided learning collaborative model to improve

implementation of EBPs in a state children's mental health system: A pilot study. *Administration and Policy in Mental Health and Mental Health Services Research*, *43*(6), 978–990.

23. Niebert, K., Marsch, S., & Treagust, D. F. (2012). Understanding needs embodiment: A theory-guided reanalysis of the role of metaphors and analogies in understanding science. *Science Education*, *96*(5), 849–877.
24. Reichstein, M., Camps-Valls, G., Stevens, B., Jung, M., Denzler, J., & Carvalhais, N. (2019). Deep learning and process understanding for data-driven Earth system science. *Nature*, *566*(7743), 195–204.
25. Samulowitz, A., Gremyr, I., Eriksson, E., & Hensing, G. (2018). "Brave men" and "emotional women": A theory-guided literature review on gender bias in health care and gendered norms towards patients with chronic pain. *Pain Research and Management*, 1, 1–*20*.
26. Schmidt, H. G., Dolmans, D., Gijselaers, W. H., & Des Marchais, J. E. (1995). Theory-guided design of a rating scale for course evaluation in problem-based curricula. *Teaching and Learning in Medicine: An International Journal*, *7*(2), 82–91.
27. Sheikh, R., & Jahirabadkar, S. (2018). An Insight into Theory-Guided Climate Data Science—A Literature Review. In *Advances in Data and Information Sciences* (Vol.1, pp. 115–125). Springer, Singapore.
28. Shi, X., & Levy, S. (2005). A theory-guided approach to library services assessment. *College & Research Libraries*, *66*(3), 266–277.
29. Squire, K. D., & Jan, M. (2007). Mad City Mystery: Developing scientific argumentation skills with a place-based augmented reality game on handheld computers. *Journal of Science Education and Technology*, *16*(1), 5–29.
30. Sullivan, P. A., Rommel, H., Liao, Y., Olbricht, B. C., Akelaitis, A. J., Firestone, K. A., ... & Eichinger, B. E. (2007). Theory-guided design and synthesis of multichromophore dendrimers: An analysis of the electro-optic effect. *Journal of the American Chemical Society*, *129*(24), 7523–7530.
31. Sun, J., Niu, Z., Innanen, K. A., Li, J., & Trad, D. O. (2020). A theory-guided deep-learning formulation and optimization of seismic waveform inversion. *Geophysics*, *85*(2), R87-R99.
32. Swan, M. (2013). The quantified self: Fundamental disruption in big data science and biological discovery. *Big data*, *1*(2), 85–99.

33. Van Der Aalst, W. (2016). Data science in action. In *Process Mining* (Vol.1, pp. 3–23). Springer, Berlin, Heidelberg.
34. VanderPlas, J., Connolly, A. J., Ivezić, Ž., & Gray, A. (2012, October). Introduction to astroML: Machine learning for astrophysics. In *2012 Conference on Intelligent Data Understanding* (Vol.1, pp. 47–54). IEEE.
35. Varshney, K. R., & Alemzadeh, H. (2017). On the safety of machine learning: Cyber-physical systems, decision sciences, and data products. *Big Data*, *5*(3), 246–255.
36. Wagner, N., & Rondinelli, J. M. (2016). Theory-guided machine learning in materials science. *Frontiers in Materials*, *3*, 20–28.
37. Way, M. J., Scargle, J. D., Ali, K. M., & Srivastava, A. N. (Eds.). (2012). *Advances in Machine Learning and Data Mining For Astronomy, 1, 1–33*. CRC Press.
38. Zahedi, F. M., Walia, N., & Jain, H. (2016). Augmented virtual doctor office: theory-based design and assessment. *Journal of Management Information Systems*, *33*(3), 776–808.
39. Zheng, X., Ji, Y., Tang, J., Wang, J., Liu, B., Steinrück, H. G., ... & Cui, Y. (2019). Theory-guided Sn/Cu alloying for efficient CO_2 electroreduction at low overpotentials. *Nature Catalysis*, *2*(1), 55–61.
40. Zimmerman, A. S. (2008). New knowledge from old data: The role of standards in the sharing and reuse of ecological data. *Science, Technology, & Human Values*, *33*(5), 631–652.

Chapter 3

Statistical Physics and Machine Learning

CONTENTS

3.1. INTRODUCTION

Fields like mathematical finance, statistics, and approximate computation have been permeated by the methodology of statistical mechanics. Similarly, significant breakthroughs have been experienced in the disciple itself, especially in the interpretation of dynamic non-equilibrium procedures. To reevaluate statistical learning as a vigorous process the tools of contemporary statistical mechanics have been used in this chapter. Moreover, the applications of modern statistical mechanisms to regulate fluctuation examination of microtubule motion in motility analysis and enzyme motion have been explained (Yuille, 1990; Malzahn & Opper, 2005).

It has become possible for the academia and industry to use machine learning in increasingly beneficial ways over the past decade due to the exponential rise in the accessibility of computational resources and massive, large datasets. By the nineties, the parallels between machine learning algorithms and statistical mechanics and were well-established (Parrondo et al., 2015; Van Der Aalst, 2016). Howsoever, the evolution of machine learning as a field of applied engineering together with the latest breakthroughs in the perception of non-equilibrium thermodynamic insists through information theory concepts demands a reevaluation of the links between the two subjects. In this chapter, the relationship between statistical learning and statistical mechanical systems will be demonstrated along with suggestions through which the theoretical outcomes from statistical mechanics can be enabled for the enhancement of the unification of learning algorithms (Carrasquilla & Melko, 2017; Gorban & Tyukin, 2018).

3.2. BACKGROUND AND IMPORTANCE

Obtaining knowledge through data about a concept or system which was previously not known is called statistical learning. Some basic problems of statistical learning are density estimation, classification, clustering, and regression (Friedman et al., 2001; Kecman, 2001). An exceptional dimensionality of either the parameter space or data individually or both of them together happens to be a common factor among the most efficient machine learning applications. The noise is a negative outcome of the availability of the massive amount of data (Wolpert, 2006; Cocco et al., 2018). The existence of randomness and increased dimensionality, are the two structural characteristics of statistical learning which are also vital in statistical mechanics, proposing that the instruments of statistical mechanics would supply a strong theoretical foundation for learning algorithms.

Considerable links between statistical mechanics, inference, and machine learning have already been reviewed (Watkin et al., 1993; Biehl & Caticha, 2003; Zdeborová & Krzakala, 2016).

At the same time, the latest discoveries at the interface of statistical physics and information theory brought to light the basic attributes of irreversibility unlike equilibrium and extended those developments to procedures consisting of information exchange (Jarzynski, 1997; Crooks, 1999). Through master equation formalism, they have presented a clear arrangement for non-equilibrium procedures (Altaner, 2014), and extended those principles to any operation that can be explained by utilizing a Bayesian network (Ito & Sagawa, 2016). This area of research is normally called information thermodynamics, produces tighter bounds as compared to traditional techniques for procedures that work far from equilibrium in locations where thermal fluctuations are applicable (Yuilleet al., 1994; Roudi et al., 2009; Baudot, 2019).

3.3. LEARNING AS A THERMODYNAMIC RELAXATION PROCESS AND STOCHASTIC GRADIENT LANGEVIN DYNAMICS

At various multiple levels, connections between statistical mechanics and statistical learning can be made. In this chapter, the learning process is modeled as the relaxation from a non-equilibrium initial state, to an equilibrium state (Gomez-Marin et al., 2008; Mestres et al., 2014). The parameters of the model were used to set a preliminary probability distribution. Consequently, this probability distribution is adjusted to reduce the empirical loss (or energy). The empirical loss is decreased by the last parameter distribution, equal to statistical mechanics' equilibrium distribution in (García-Palacios & Lázaro, 1998). This dynamic learning arrangement is framed as a route on the high dimensional statistical manifold, a Riemannian manifold that's metric is presented to be the Fisher information matrix (Regueraet al., 2005; Amari & Nagaoka, 2007).

Numerous modifiable algorithms for optimization malfunction due to a single or maximum a posteriori (MAP) solution. Insertion of thermal noise is crucial to prevent the posterior from collapsing. Originally introduced in Welling and Teh (2011), these kinds of methods are known as Stochastic Gradient Langevin Dynamics (SGLD). In the following chapter, this method and its variants, improvements, and possible drawbacks will be analyzed (Leimkuhler et al., 2019).

3.4. CHEMOTAXIS IN ENZYME CASCADES

As a reaction to the initial enzyme's substrate, enzymes involved in the reaction cascades have been seen to form multi-enzyme structures, also called metabolons (Zhao et al., 2013). Experimental evidence related to directed chemotactic activity of enzymes in regards to their substrate gradients will be presented, and a theoretical diffusion model will be suggested to explain the occurrence of the purine synthesis cascade (Krem& Di Cera, 2002; Jørgensen et al., 2005). The following purinosome, or metabolon, has been observed experimentally in the past. It would be possible to develop new treatments of purine synthesis disorders that focus on the purinosomes or the components initiating its assembly (Bray et al., 1993; Zhao et al., 2018).

3.4.1. Meaning and Background

The relationship between enzymes in living cells is one of the most actively researched areas. Substrate channeling seems to be supported by the production of metabolons due to the presence of the first substrate (An et al., 2008; Castellana et al., 2014; Wu & Minteer, 2015). By regulating reaction intermediates along a certain course from one enzyme to the next, substrate channeling aids sequential reactions with high selectivity and output. There is a rise in the diffusive motion of enzymes as a function of reaction rate and substrate concentration (Muddana et al., 2010; Sengupta et al., 2013; Riedel et al., 2015). The proposed evidence suggests that a mechanism of sequential, directed chemotactic movement leads to the association of enzymes through the purine synthesis metabolic pathway in which the result of one is the substrate, for the next. These steps in living cells may lead to the production of metabolons encompassing the mitochondria, which is a source of ATP (Butler et al., 2015; French et al., 2016).

A phenomenon known as cross-diffusion or self-diffusiophoresis includes the generation of areas of higher enzyme concentrations which develop as a result of diffusion of an enzyme up the gradient of its substrate as depicted in the experimental evidence. This phenomenon has been analyzed in both experimental and theoretical studies (Vergara et al., 2002; Annunziata et al., 2009; Vanag & Epstein, 2009).

In addition to giving an understanding of how the cells utilize spatial control to maintain enzymes and enzyme complexes and boost metabolic efficiency, metabolon formation can also be possibly explained by the fundamental process of purinosome production. Finer drugs can be manufactured to target metabolic diseases through a better understanding of these processes (Paduano et al., 1998; Fu et al., 2015).

3.4.2. Experimental Design

By utilizing photolithography the collaborators synthesized a microfluidic flow device to analyze the activity of enzymes in a cascade towards a substrate gradient. Specific thiol-reactive Dylight and amine-reactive dyes were used to fluorescently label the first enzyme, aldolase (Ald) and the fourth enzyme, hexokinase (HK) of the glycolytic cascade. Confocal microscopy was used to evaluate the movement of these enzymes over the channel (Bouchet et al., 2010; Aurell et al., 2012).

3.4.3. Cross Diffusion Model

It has been suggested that the cross-diffusion effects are the reason behind the chemotactic aggregation of enzymes in areas of high substrate concentrations. The Fickian diffusion of enzymes which includes enzyme transfer from areas with higher enzyme concentration to those with lower enzyme concentration is canceled out by cross-diffusion as it causes substrate-gradient induced aggregation (Postma& Van Haastert, 2001). During the availability of its substrate cross-diffusion differs from an enzyme enhanced diffusion, which is also checked for stable substrate concentrations and escalates the equilibrium of the concentration of enzymes by Fickian diffusion. An overall theoretical detail of diffusion in a multi-element system integrates the course of the same species against the concentration gradients of other species present in the solution and the course of the same species relative to its concentration gradient (Fick's law). In the availability of its substrate S, the diffusive motion for the concentration c_e of unbound enzyme E can be stated as:

$$J_e = -D\Delta c_e - D_{XD}\Delta c_s \qquad (1)$$

Where Δc_s and Δc_e are gradients in substrate and enzyme concentrations, D is the Fick's law diffusion coefficient and D_{XD} is the "cross-diffusion" coefficient, respectively. It will be demonstrated that the model describes the increase in the development of metabolons in enzyme cascades and the transfer of enzymes up their substrate gradient.

3.5. ACQUIRE NANOSCALE INFORMATION THROUGH MICROSCALE MEASUREMENTS UTILIZING MOTILITY ASSAYS

Understanding systems like engineered nanodevices, cellular cargo transport, cell division, and muscle contraction is crucial for assessing the behavior of coupled molecular motors. By using the gliding motility assay as a model approach a demonstration through a collection of Brownian dynamics simulations and experimental data was carried out. It showed that fluctuation examination of the microscale microtubule motion can be used to obtain quantitative outcomes about the behavior of the nanoscale motors. Determining the heterogeneity in motor force production by factor α is particularly intriguing. Experimental data will be used to compare the theoretical output of the model. The conclusions on the proficiency of an idea of an experimental approach to assess nanoscale dynamics utilizing observed microscale assembled motor protein action will also appear as proof of fluctuations (Carlsson, 2010).

3.6. STATISTICAL LEARNING

Statistical physics strives to classify the probabilistic laws conducting the equilibrium state of a common system with multiple degrees of freedom. The goal of parametric statistical learning is to grasp a rule by reducing an objective function concerning a highly structural parameter vector.

A mechanical system will progress towards the reduction of its energy. Negligible energy for a constant temperature, the microscopic foundation of the system will develop into its equilibrium distribution, which will have the maximal entropy for the given energy.

A basic overview of the ties between statistical physics and statistical learning will be provided in this section, especially focusing on geometric concepts and simulation techniques. Terms and notations used to represent the ties between statistical learning and statistical mechanics will be presented briefly in this chapter.

3.6.1. Equilibrium Statistical Mechanics

The differential first law is stated as:

$$dE = -pdV + TdS - \sum_{i} u_i dN_i \qquad (2)$$

where μ_i and N_i is the chemical potential and number of species I, E is the energy of the system being discussed, T the temperature, P the pressure, S the entropy, and V is the volume.

3.6.1.1. Selection of An Ensemble

Experimental conditions govern the selection of the ensemble in thermodynamics. The definition of a suitable ensemble is more complicated in statistical learning. The canonical ensemble will be taken as the "correct" ensemble.

3.6.1.2. Boltzmann Distribution

The derivation from Chandler (1987) will be observed. The system under discussion will be assumed to be in the canonical ensemble, it is therefore kept at a stable temperature T by exposure with a huge heat bath R. For the system, two different states, s_1 and s_2, will be considered. In these two states, the quantity of configurations accessible to the reservoir and system is equivalent to the amount of configurations $\Omega_R(s_1)$ and $\Omega_R(s_2)$ feasible to the reservoir only (as a specific condition for the system has been chosen).

In that specified state, the probability of the system is consistent with the number of microstates of the reservoir for that particular state, so it can be stated as:

$$\frac{p^*(s_1)}{p^*(s_2)} = \frac{\Omega_R(s_1)}{\Omega_R(s_2)} \qquad (3)$$

$S(s_i) = -k \ln\Omega_R(s_i)$ is also present, so:

$$\frac{p^*(s_1)}{p^*(s_2)} = \exp\frac{1}{k}(S_R(s_1) - S_R(s_2)) \qquad (4)$$

Eventually, Eq. (2) can be combined at a uniform volume, and temperature among these two states: $\Delta S_R = 1/T\Delta E_R$, as both s_1 and s_2 are equilibrium states. It is known that the total energy is preserved so $\Delta E_R = -\Delta E_s$:

$$\frac{p^*(s_1)}{p^*(s_2)} = \frac{e^{-E(S_1)/kT}}{e^{-E(S_2)/kT}} \qquad (5)$$

The Boltzmann probability distribution p^* is written in the canonical ensemble considering that it is acceptable for any two states:

$$p^*(s) = \frac{e^{-E(s)/kT}}{Z} \qquad (6)$$

where $Z = \sum_s \exp\{-E(s)/kT\}$ is the partition funaction

Normally in a system with a clearly defined energy function, but a great number of degrees of freedom, it is easy to calculate the numerator but the partition function is mostly intractable.

3.6.2. Microscopic Thermodynamic Quantities, and Additional Insight into the Boltzmann Distribution

The energy E is a highly specific scalar in macroscopic thermodynamics. As a probabilistic system is being presumed, the quantity as a mean under the canonical distribution p^* can be deduced as:

$$E = \int_s \in (s) p^*(s) ds \qquad (7)$$

In a given microstate s, ε(S) is the energy of the system. The function will be interpreted as constant and associated with the settings of the experiment, however with a driving parameter λ in a driven arrangement; the function will change during the experiment as it will be dependent on λ.

Entropy can be written, in a microscopic view:

$$S(p^*) = -\int_s p^*(s) \ln p^*(s) ds \qquad (8)$$

Throughout this chapter, Boltzmann's constant k will be considered as a unity, as shown in the above equation. The fact that p^* is the probability distribution, that for the specified E enhances S can conveniently be shown is important for the understanding of the non-equilibrium dynamics of learning.

3.7. NON-EQUILIBRIUM STATISTICAL PHYSICS

Major inroads concerning the theoretical knowledge of non-equilibrium procedures in statistical physics have been made in the past two decades. During a non-equilibrium procedure, Crooks (1999a; 1999b) and Jarzynski(1997) associated the fluctuation theorems with equilibrium free-energy dissimilarities. Interest in both experimental and theoretical non-equilibrium statistical physics has been reignited due to these theorems (Collin et al., 2005; Bérut et al., 2012; Lu et al., 2014).

The requisite to theoretical apprehension of non-equilibrium states depends on how much more or less, data the non-equilibrium state has in contrast to the parallel equilibrium state. This notion will be more clearly defined in the following paragraphs (Muddana et al., 2010; England, 2013; Frey, 2013).

3.7.1. Shannon Information

The word information denotes how much is understood about a random variable when it is sampled from, in information theory (Cover & Thomas, 2012) For instance, sampling from a consistent random variable never serves at providing any information. Meanwhile, a specific amount of information is obtained after tossing a coin. One bit (1 for tails and 0 for heads) can be used to indicate the result of that coin toss.

$\text{Log}_2 8 = 3$ bits of information are required to symbolize the outcome of a fair if an 8-sided die is considered. By utilizing shorter representations for the outcomes with more potential, the length of the average description may be reduced if the die was unfair (Sagawa & Ueda, 2010; Deffner& Lutz, 2012).

In non-equilibrium thermodynamics, the idea of information as the average number of bits that are required to denote the outcome of a random variable is very fundamental.

This Shannon information or the average quantity of information is very similar to Eq. (8):

$$H(p) = -\sum_i p(i)\log p(i) \qquad (9)$$

3.7.2. Non-Equilibrium Thermodynamic Quantities

The system has a detailed Hamiltonian and equilibrium probability distribution p*, in this approach. In a survey of a non-equilibrium scenario when the distribution of the system is originally out of equilibrium and houses a probability distribution p 6= p*.

Non-equilibrium energy can be stated as:

$$E(p) = \int p(s) \in (s)ds \qquad (10)$$

And broaden the terminology of entropy to this non-equilibrium probability:

$$S(p) = \int p(s)\log p(s)ds \qquad (11)$$

3.7.3. Relaxation from a Preliminary Non-Equilibrium State

The topic of non-equilibrium statistical mechanics will be limited to the point where a system is originally not in equilibrium in terms of its Hamiltonian, then renders into its equilibrium state. This will coincide with the statistical learning process.

As viewed in the earlier sections, non-equilibrium probability distribution p is what describes the non-equilibrium state. The Kullback-Leibler Divergence $D_{KL}(p,p^*)$, in information theory, is the supplementary number of bits required to encode the result of a random variable succeeding p when the encoding process was improved for a random variable succeeding p^* (Cover & Thomas, 2012):

$$D_{KL}(p,p^*) = \int p(x)\ln\frac{p(x)}{p^*(x)}ds \tag{12}$$

It can be illustrated that D_{KL} is zero only when $p = p^*$, it is constantly positive (by Jensen's inequality), even though it is not a distance (it is unsymmetrical).

The relaxation dynamics are now being identified as the dynamics of time-dependent probability distribution p_t from an introductory non-equilibrium state p_0 to the equilibrium state $p_\infty = p^*$. By following (Altaner, 2017) and describing a weakly relaxing dynamics as a sequence so that

$$\lim_{x\to\infty} D_{KL}(p,p^*) = 0 \tag{13}$$

However, strongly relaxing dynamics is consistently discarding information related to the encoding explained by the equilibrium probability distribution:

$$\frac{\partial D_{KL}(p_t,p^*)}{\partial t} \leq 0 \tag{14}$$

It can be depicted that any Markovian memoryless dynamics concentrating to an equilibrium distribution is highly relaxing. Hence Langevin dynamics define a strongly relaxing system, and it can be presumed that the relaxation dynamics are strongly relaxing (Chan et al., 2014).

3.7.4. The Non-Equilibrium Second Law, Entropy Production, and Entropy

The entropy of a remote system can only rise as claimed by the second law. For the system s-linked with a reservoir, it can be stated that $\Delta S_{tot} = \Delta S_R + \Delta S_s \geq 0$. Additionally, heat exchanges at consistent T are the only transitions in entropy in contact with the reservoir, so $\Delta S_R = Q/T$, where Q is the heat from the complex to the reservoir. Hence, it can be stated that:

$$\Delta S_s = \Delta S_s^{exchange} + \Delta S_s^{irr} \quad (15)$$

With $\Delta S_s^{exchange} = -Q/T$. Therefore, the second law $\Delta S_{tot} \geq 0$ can then be restated as:

$$\Delta S_s^{\pm} \geq 0 \quad (16)$$

Under the presumption of strongly relaxing dynamics, it can be seen that the unalterable entropy formation is correspondingly the time derivative of the KL divergence among the corresponding equilibrium distribution and the non-equilibrium probability distribution:

$$\frac{\partial S^{irr}}{\partial t} - -\frac{\partial D(p_t, p^*)}{\partial t} \leq 0 \quad (17)$$

It can be observed that strongly relaxing dynamics always discards information (information here can be perceived as the average description length of the scenario when the encoding was expanded for the equilibrium distribution).

A question that can normally be formulated from the perspective of trajectory in phase space where the KL deviation is minimized at every stage is: What is the phase space's ideal trajectory that will guide equilibrium distribution? The principal notions of differential geometry need to be formed to answer this question (Soliman et al., 2010; Feng et al., 2014).

3.8. PRIMARY DIFFERENTIAL GEOMETRY

As mentioned above, equilibrium distribution is gained after the relaxation of a non-equilibrium state that can be viewed as a trajectory in a space of parameterized probability distributions p_t. In which parameters describe the positions (as well as the velocities) of the entire particles in the system (Manasse&Misner, 1963; Ciarlet, 2005). To a certain extent, a more general framework will be considered of a space of parameterized probability distributions p_t parameterized by any parameter set θ.

3.8.1. Motivating Example

Parameterized by its standard deviation σ and mean μ, the space of Gaussian probability distributions can be reviewed to analyze why notions from differential geometry are required to examine a space of probability distributions. Initial judgment would lead towards the consideration of space to be like R^2 supplied with the traditional Euclidian distance. Hence with this metric, the distance between N (0,0.1) and N(1,0.1) is naturally $(0 - 1)^2 + (.1 - 0.1)^2 = 1$. While the distance between N (0,1000) and N(1,1000) is the same, $(0 - 1)^2 + (1000 - 1000)^2 = 1$.

It is now clear what is the problem with the Euclidian distance. The two low-variance Gaussians are extremely distinct from each other. Their mean has a gap of more than ten standard deviations, and it would be highly odd to mistake a sample from one of the low-variance Gaussians with a sample from the other. On the other hand, the high variance Gaussians are very identical, and it would be very difficult to differentiate samples from the two distributions (Novikov & Fomenko, 2013).

Introduction of a better notion of commonality between distributions in equation 12: the KL divergence. The KL-divergence although cannot be utilized as a distance since it is not symmetric. The application of symmetrized KL divergence $\tilde{D}$ can be used to solve the problem (Amari & Nagaoka, 2007):

$$\tilde{D}_{KL}(p, p^*) = \frac{D_{KL}(p, p^*) + D_{KL}(p^*, p)}{2} \tag{18}$$

For 1-dimensional Gaussians, the 2-dimensional space can be written as:

$$\tilde{D}_{KL}(N(\mu,\sigma), N(\mu^*,\sigma^*)) = \frac{1}{4\sigma^2\sigma^{*2}}\left[(\mu^* - \mu)^2(\sigma^2 + \sigma^{*2}) + (\sigma^2 - \sigma^{*2})^2\right] \tag{19}$$

Whereas, $0.5.10^{-6}$ can be estimated to be the distance between the two high-variance Gaussians, while 50 can be computed to be the distance among the two low-variance Gaussian of the motivating example, which is much more synchronized with perceptivity.

It can be observed that the space of one-dimensional Gaussian distributions is curved: the level of standard deviation determines the distance between distributions containing two means divided by the same interval. Symmetrized KL divergence Eq. (16) is a common method to explain distance in that space. To evaluate how routes between distributions can be enhanced in such spaces, a step towards the direction of formalization of these approaches will be taken (Lavendhomme, 2013).

3.8.2. Fisher Information

The Fisher Information in classical statistics revolves around the estimation of Cramer-Rao regulated by the variance of any unbiased maximum probable estimator θ̂ (Wasserman, 2013):

$$Var\ \hat{\theta} \geq \frac{1}{I(\theta)} \tag{20}$$

Hence, I(θ) is the Fisher Information, commonly defined as:

$$I(\theta) = \mathbb{E}\left[\left(\frac{\partial.}{\partial\theta}\log f(X;\theta\right)^2 |\theta\right] \tag{21}$$

where f is the possibility of the data X stated parameters θ.
It can be demonstrated that the Hessian of the symmetrized KL divergence is also the Fisher Information matrix. Therefore curvature in the space of parametrized probability distributions can be specified by the Fisher Information:

$$I(\theta) = \nabla_\theta^2 D_{KL}(p_\theta, p_{\theta^*})\Big|_{\theta=\theta^*} \tag{22}$$

3.8.3. The Distance on a Riemannian Manifold

Until now it has been detected that the space of probability distributions happens to be curved. It can be revealed (Amari & Nagaoka, 2007) that it is a Riemannian manifold that may be simply defined as a curved space that limitedly resembles R^n. For instance, even though each point is locally similar to R^2, a sphere in R^3 is curved.

On a space like this, notions of distances, dot products, and angles are all local: distances in patches with lower curvature are not similar to distances in a patch with higher curvature.

Entire notions of distances and angles rely on the dot product of two particular vectors <**u**,**v**> in Euclidean spaces. While in a Riemannian manifold, this vector product at a point θ is revised for the curvature utilizing the metric tensor **F**: <**u**, **Fv**>, and a locally explained length as $\|u\| = \sqrt{<u, Fu>}$.

Here p_θ and p_θ*, the distance between these two points is thus a geodesic, the minimal curve length of paths between these two points, provided the curve length along a path λ(t) so that λ(T) = p_θ* and λ(0) = p_θ is determined as:

$$l_\lambda(\theta,\theta^*) = \int_0^T \left\|\lambda'(t)\right\| dt \quad (23)$$

Where the Riemannian metric tensor mentioned above is used to explain the norm ||.||. Hence, the distance is:

$$d(p_\theta, p_{\theta^*}) \min_\lambda l_\lambda(\theta,\theta^*) \quad (24)$$

The relevant metric **F** is the Fisher information matrix for the space of probability distributions:

$$F_\theta = I(\theta) \quad (25)$$

Even after the introduction of the notion of distance on a parametric probability space utilizing the symmetrized KL divergence, and that the local curvature of the space is the second result of this distance, the distance among two points on the probability manifold is not completely equivalent to the symmetrized KL divergence.

3.8.4. The Natural Gradient

An effective technique to detect parameterized probability distribution's minima of a function is the Natural Gradient method which was first suggested to be applied to neural networks in Amari and Nagaoka (2007). Taking the example of a non-equilibrium relaxation process again and beginning with the preliminary probability distribution p while relaxing into the equilibrium probability distribution p* which reduces the system's energy. The distance between p* and p is the smallest path length between the two distributions. Finding an ideal path p_t would be expensive numerically, particularly for a large number of particles.

The accepted method to measure an optimal path locally would be to utilize the steepest descent at each step: quantify the local gradient, and then form a step in that direction:

$$\theta_{t+\Delta t} = \theta_t + \lambda \nabla_\theta E(\theta) \quad (26)$$

The small step size is symbolized by λ.

Despite that, the above equation only explains the steepest descent method in a Euclidean space. Keeping the curvature in mind, the data requires changes as the probability space is curved.

What is the direction that will strengthen the transition in E? This can be formalized as a maximization problem:

$$\max E(\theta+\delta\theta)u.c.\|\delta\theta\| < \in \quad (27)$$

That can be re-stated as:

$$\max E(\theta) + \in v^T \nabla E(\theta) u.c.\langle v, F_\theta v\rangle = 1 \quad (28)$$

The Natural Gradient method is obtained in a result when the equation is solved with the conventional technique of Lagrange multipliers:

$$\theta t + \Delta t = \theta t + \lambda \mathbf{F} - \theta\ 1 \nabla \theta E(\theta) \quad (29)$$

In the case of a simple Euclidean space, it has been observed that the Riemannian metric tensor is the identity, and the classic steepest descent method is retrieved.

3.8.5. Geodesics on Thermodynamic Length and the Probability Manifold

An association between thermodynamic length and fluctuations has been made by Frey (2013). It suggests that a geodesic on a Riemannian manifold is a distance between two equilibrium distributions which can also be assumed as the number of natural fluctuations on the finite-time path among the two distributions. Frey (2013) also calculated the relative Riemannian metric and proposed that it is equal to the Fisher information metric – hence the actual Riemaniann manifold for thermodynamics is the statistical Riemannian manifold which has been shown above.

3.9. BAYESIAN MACHINE LEARNING AND CONNECTIONS TO STATISTICAL PHYSICS

Parametrized supervised learning; a subset of machine learning will be introduced in this section. It includes a wide subset of machine learning therefore to present the limitations of computational approaches and to fix concepts; examples in deep neural networks will be given (Wolpert, 2006; Friston, 2012; Zdeborová & Krzakala, 2016).

3.9.1. Learning Formalism

The formalism has been selected as it can easily be generalized to multiple types of supervised learning. In the most conventional formalism possible, a matrix consisting of observed variables $\mathbf{X} \in \mathbf{R}^{n\times p}$, where p is the dimension of the observed variables, and n is the number of observations. For instance, the $1\times p$ row vectors would be some ciphering of the pixel values of the image and **X** could represent be a set of n images denoting animals. There is also a vector $\mathbf{y} \in \mathbf{R}^n$ incorporating the labels for every example (namely, a number expressing which animal is in which picture).

Statistical learning aims to discover a function $f(\mathbf{X},\theta)$ of the input **X**, which is parameterized by a vector θ whose output comes close to **y**. As the infinite-dimensional case embodies a non-parametric estimation, the dimension of θ has not been defined.

Every example (X,y) is a random variable formed from an unknown probability distribution, in probabilistic terms. The target is then to reduce the risk function:

$$R(f,X,y,\theta) = \mathbf{E}[L(f(X,\theta),y)] \qquad (30)$$

Where **E** is the expectation regarding the true joint distribution of (X,y) and L is some presumed loss function (for instance the L_2 norm of $f(X)-y$).

However, there is no access to his joint distribution. Thus, the center of attention of statistical learning turns to be the minimization concerning θ of the empirical risk Rˆ; which is the ERM, Empirical Risk Minimization principle (Vapnik, 1999; Wasserman, 2013):

$$R\hat{}(f,\mathbf{X},\mathbf{y},\theta) = hL(f(\mathbf{X},\theta),\mathbf{y})i \qquad (31)$$

The brackets in the equation signify the average overall examples. Understanding from examples or reducing the empirical risk is usually done utilizing gradient descent (calculating the gradient complete examples at the same time, then modifying the parameters θ appropriately) or more frequently the stochastic gradient descent (estimating the gradient and altering the parameters consequently for each example).

The ERM minimization principle is not ideal, as viewed, even in the case of large numbers of examples. If the number of parameters is extremely great, consistently larger than the number of examples, the outcomes can normally be memorized by the learning function. That can generate an empirical risk of zero however; it does not assure favorable performance on out of sample examples (overfitting).

Characterized by layers of "neurons," neural networks are a particular class of functions f, which is commonly a linear modification of the past output layer accompanied by a non-linear "activation function" σ. The hyperbolic tangent or the sigmoid is the prevalent example of activation functions. For layers i= 0,1,...,I, it can inscribed as:

$\mathbf{O}_{i+1} = \sigma(\mathbf{W}_i\mathbf{O}_i)$ (32)

where $\mathbf{O}_I$ is the output that is compared to y, and $\mathbf{O}_i$ is the output of layer i, $\mathbf{O}_0 = \mathbf{X}$. The matrices $\mathbf{W}_i$ are of dimension $h_{i-1} \times h_i$ and signify the weights of the neural network, where h_i represents the number of neurons on layer i. It should be observed that it is normal to add an additive bias to every layer, which is excluded here for conciseness.

3.9.2. Stochastic Gradient Descent for Empirical Risk Minimization

The loss of function L $(y_i|\mathbf{x}_i;\theta_i)$ has been specified and minimized concerning θ. A regularizing term R (θ) is incorporated into the minimization problem deliberately. The problem of minimization can thus be stated as:

$$\hat{\theta} = \arg\max \sum_i -L(y_i, x_i;\theta) + R(\theta) \quad (33)$$

For deep neural networks, this minimization is achieved by employing Minibatch Stochastic Gradient Descent.

θt+Δt = θt+ λt∇LMB(t) (34)

where $L_{MB(t)}$ is the loss calculated solely on a random subset (or minibatch) of the details and λ_t is a declining function to $\sum \lambda_t = +\infty$ and $^{P}{}_{\infty}\lambda^2_t < +\infty$.

To reduce the loss of deep neural networks, this technique has proven to be very effective. Additionally, as it does not need to fit the complete dataset in memory, it scales effortlessly.

3.9.3. Bayesian or Probabilistic Neural Networks

A method in the above segment has been outlined to gauge the vector of parameters θ that adequately minimizes the loss L on the data **(X,y)**. Containing a point estimate is acceptable in a majority of applications. Nevertheless, it is most favorable to measure a probability distribution on the parameter vector θ, as it would deliver (1) a method to possibly identify out-of-sample data, (2) a certain regularization scheme, and (3) a way to assess the error on the predictions.

3.9.4. Bayesian Estimation

If the model currently being used is defined as a parametric conditional probability function p((y, X)|θ), the Bayes' theorem can be used to write:

$$p((y,X)|\theta)p(\theta) = p(\theta|(y,X))p((y,X)) \quad (35)$$

where p (θ|(y, X)) is the posterior probability of the parameters provided in the data, p(θ) is the prior probability of the parameters (mostly selected to be uninformative), and p((y, X)|θ), viewed as a function of θ is the probability. Together with the probabilistic model and a guideline of the priors on the parameters, permits the estimation of the probability, and thus provides the posterior on θ. The target distribution, the posterior, extends the ability to analyze errors over the parameters; however, it also employs the model for prediction by receiving expectations of the probability over the parameters utilizing the posterior.

3.9.5. Loss as Energy

According to the Boltzmann distribution a type of the probabilistic model attached to the parameters has been hypothesized and specified in equation 6, where the energy is restored by the loss function L:

$$p(\theta|(X,y)) = (\exp[-L(\theta)])/Z \qquad (36)$$

where $Z = \int \exp[-L(\theta)]d\theta$ normalizes the expression to a probability.

If each of the loss terms is altered and regularized term as energies, with Gibbs probability distributions p(E) = (exp−E)/Z, the above equation can be taken as the MAP estimate of the succeeding probabilistic model:

$$p(\theta|x) = p(\theta)\prod_i p(y_i, x_i|\theta) \qquad (37)$$

where $\log p(y_i, \mathbf{x}_i|\theta) = L(y_i, x_i;\theta)$ is the log-probability, $\log p(\theta) = R(\theta)$ is the log-prior, and $p(\theta|\mathbf{x})$ is the posterior probability of the parameters.

3.10. STATISTICAL PHYSICS OF LEARNING IN DYNAMIC PROCEDURES

A non-equilibrium relaxation procedure in thermodynamics can be explained as a restructuring from a primary probability distribution p to a closing Boltzmann equilibrium probability distribution p^*. A strongly relaxing reformation eliminates information constantly, thereby decreasing the distance between p and p^* frequently. The prior probability of the parameters θ expresses a basic distribution p on these parameters, in statistical learning. These parameters are modified gradually towards the estimation of the minimum empirical loss distribution p^* through a sequential optimization scheme (Bodmer et al., 2004; Castellano et al., 2009).

3.11. EARLIER WORK

The last two decades have witnessed multiple accomplished attempts to employ statistical mechanical tools and ideas in machine learning. Those connections have been investigated below.

3.11.1. Monte Carlo Evaluation of High Dimensional Integrals

The pivotal quantity in equilibrium statistical physics is the partition function $Z = \int \exp\left[-\frac{E(s)}{T}\right] dS$. The state vector s is without exception high-dimensional, and the examination of this integral is non-essential. Saddle point approximations are adopted (Chandler, 1987) in ordinary cases, however, in the majority of practical settings, these integrals are analyzed through Monte Carlo approximations. This evaluation issue is identical to the estimation of posteriors in probabilistic models. Surely, when provided with a probabilistic model, it is challenging to evaluate integrals but uncomplicated to sample from the posterior as the parameter space is mostly ten-to a hundred-dimensional (Robbins & Monro, 1951; Gelman et al., 2014).

By concentrating on sampling in high-probability regions, the Markov Chain Monte Carlo approximations strive to proficiently sample from the state space, which is done achieved using a Markov "walker." The successive step is suggested at random in the state space, and if the proposed state's energy is lower, it is highly likely to be accepted. The algorithm, the Metropolis-Hastings (1953) sampling scheme, can get obstructed in local energy minima and occasionally fail to efficiently sample the state space. In addition to that, as the walker only steps locally, its mixing times are normally slow (Goldt & Seifert, 2017).

Hamiltonian Monte-Carlo, a physics-inspired, is a remarkably more effective sampling scheme, originally discovered by physicists as "Hybrid Monte Carlo" and later on revived and publicized in the mathematical statistics body by Radford Neal (2011). Hamiltonian Monte Carlo (HMC) centrally focuses on sampling from the state space, including sample momenta, and to utilize the momenta to present the next stage, permitting bigger jumps and more systematic research of state space. An extremely detailed exploration of HMC can be viewed in Betancourt (2017).

3.11.2. Combinatorial Statistical Physics of Learning

Statistical physics of learning, when implored through combinatorial approach (Van den Broeck, 2013), intends to quantify generalization curves for machine learning issues when provided with a learning rule, for example, the Hebbian rule, and a model like a perceptron (Linsker, 1988; McHale et al., 1997).

3.11.3. Learning on Networks and Spin Glasses

This approach is largely used to this day as it has been the most beneficial in connecting statistical learning and statistical physics (Mezard & Montanari, 2009; Advani et al., 2013).

The approach involves drawing links by likening learning issues to the definition of spin glasses in statistical physics. An informal explanation of the approach has been given below.

3.11.4. The Ising Model

The Ising Model exhibits a symmetrical model of interlacing spins on a lattice, in statistical physics. As claimed by the Hamiltonian, the spin at position i, $\sigma_i = \pm 1$ acts with adjacent spins j:

$$H(\sigma) = \sum_{i,j\,neighbors} J\sigma_i\sigma_j + h\sum_i \sigma_i \tag{38}$$

where h donates an external field and J ∈ R represents the constant for the interaction energy. If $J < 0$ the interaction is anti-ferromagnetic but if $J > 0$ the present interaction is ferromagnetic. In two dimensions or more it has been demonstrated that this system shows a phase transformation between an uncontrolled phase at elevated temperatures and a controlled phase at low temperature.

3.11.4.1. Spin Glasses

Spin Glasses embody an extensive case of the Ising model, in which $J = J_{ij}$, the interaction energy constant is a random variable and its values are determined at the start of an experiment. Currently, there are two degrees of randomness: the randomness caused by the fast fluctuations of the spins σ_i and the quenched randomness brought about by the coefficients J_{ij}. Interest has been sparked in the general operation of the system especially in the estimates of thermodynamic quantities averaged over the two levels of fluctuations (Mézard et al., 1987).

3.11.4.2. Association with Statistical Learning

Numerous different techniques can be used to formalize the association between statistical learning and Spin Glass models. A detailed review of these connections can be seen in Mezard and Montanari (2009). The connection formed among spin glasses and probabilistic graphical models will be summarized in the following paragraphs.

Probabilistic Graphical Models:

Probabilistic models consist of a subset known as Probabilistic Graphical Models. The model is illustrated by a graph particularizing the dependency structure among variables. A few conventional structures are Conditional Random Fields (CRFs), which include undirected graphs or, Bayesian networks directed graphs with hierarchical dependencies (Koller& Friedman, 2009). The probability of a specified state is stipulated by an energy function expressed as a function of the estimates of the nodes of every maximum clique. A technique to present a complicated probabilistic model of multiple variables in a tractable way is through probabilistic graphical models. Surely, defining the entire joint distribution for every variable included in the model is mostly impossible. Probabilistic graphical models (PGMs) permit the joints' factorization, therefore significantly decreasing the dimensionality of the issue.

After viewing details of the Ising model mentioned above, it is simple to notice that the specification of interaction energies for surrounding nodes on the lattice is equal to establishing a graphical model on a lattice. As compared to the actual probabilistic models, the Ising model is uniform. Hence, depending on the specification of the issue, probabilistic models will consist of random interaction energies (Moore, 2017).

MarkovNets.jl:

For additional research on the connection between statistical learning and statistical mechanics on conditional random fields, the library MarkovNets.jl, was created in cooperation with the Stanford Artificial Intelligence Laboratory. Since it is written with Julia, a high-level language, extension, and alteration become extremely simple. Furthermore, Julia is the sole high-level scripting scientific language to execute just in time compilation and type-inference, causing it to operate as fast as lower-level languages, like C++. The package along with its documentation can be found on https://github.com/henripal/markovnets.jl.

3.12. LEARNING AS A QUENCHED THERMODYNAMIC RELAXATION

Thermodynamic relaxation and learning can be viewed as extremely identical processes. Both of them begin with the prior on the parameters, a primary probability distribution. Then at every stage, each of these techniques centralizes to Boltzmann distribution or equilibrium through minimizing the KL divergence among the current distribution and the equilibrium distribution (Vlcek et al., 2019).

It is suggested to use a basic outcome of information thermodynamics and to practice it on statistical learning. The finest explanation for the foundational conclusion of information thermodynamics given in Parrondo et al. (2015), proposes that the relaxation from a non-equilibrium state to a state of equilibrium generates work (or negative work). When the Hamiltonian is quenched to complement the fundamental, non-equilibrium work distribution (modifying the distribution to an equilibrium distribution), then distorted quasi-statically to the chosen equilibrium distribution, the maximum absolute capacity of work would be produced by this relaxation. It has been recommended to study the performance of this process as practiced on the learning process of a neural network.

REFERENCES

1. Advani, M., Lahiri, S., & Ganguli, S. (2013). Statistical mechanics of complex neural systems and high dimensional data. Journal of Statistical Mechanics: Theory and Experiment, 2013(03), P03014.
2. Altaner, B. (2014). Foundations of stochastic thermodynamics (Vol.1, pp. 1–32). arXiv Preprint arXiv:1410.3983.
3. Altaner, B. (2017). Nonequilibrium thermodynamics and information theory: basic concepts and relaxing dynamics. Journal of Physics A: Mathematical and Theoretical, 50(45), 454001.
4. Amari, S. I., & Nagaoka, H. (2007). Methods of Information Geometry (Vol. 191, pp. 1–60). American Mathematical Soc.
5. An, S., Kumar, R., Sheets, E. D., & Benkovic, S. J. (2008). Reversible compartmentalization of de novo purine biosynthetic complexes in living cells. Science, 320(5872), 103–106.
6. Annunziata, O., Vergara, A., Paduano, L., Sartorio, R., Miller, D. G., & Albright, J. G. (2009). Quaternary Diffusion Coefficients in a Protein–Polymer–Salt–Water System Determined by Rayleigh Interferometry. The Journal of Physical Chemistry B, 113(40), 13446–13453.
7. Aurell, E., Gaw-dzki, K., Mejía-Monasterio, C., Mohayaee, R., & Muratore-Ginanneschi, P. (2012). Refined second law of thermodynamics for fast random processes. Journal of Statistical Physics, 147(3), 487–505.
8. Baudot, P. (2019). The Poincare-Shannon machine: statistical physics and machine learning aspects of information cohomology. Entropy, 21(9), 881.
9. Bérut, A., Arakelyan, A., Petrosyan, A., Ciliberto, S., Dillenschneider, R., & Lutz, E. (2012). Experimental verification of Landauer's principle linking information and thermodynamics. Nature, 483(7388), 187–189.
10. Betancourt, M. (2017). A conceptual introduction to Hamiltonian Monte Carlo (Vol.1, pp. 1–33). arXiv preprint arXiv:1701.02434.
11. Biehl, M., & Caticha, N. (2003). Statistical Mechanics of On--line Learning and Generalization. In The Handbook of Brain Theory and Neural Networks, 1, 1–22.
12. Bodemer, D., Ploetzner, R., Feuerlein, I., & Spada, H. (2004). The active integration of information during learning with dynamic and interactive visualisations. Learning and instruction, 14(3), 325–341.

13. Bouchet, F., Gupta, S., & Mukamel, D. (2010). Thermodynamics and dynamics of systems with long-range interactions. Physica A: Statistical Mechanics and Its Applications, 389(20), 4389–4405.
14. Bray, D., Bourret, R. B., & Simon, M. I. (1993). Computer simulation of the phosphorylation cascade controlling bacterial chemotaxis. Molecular Biology of the Cell, 4(5), 469–482.
15. Butler, P. J., Dey, K. K., & Sen, A. (2015). Impulsive enzymes: a new force in mechanobiology. Cellular and Molecular Bioengineering, 8(1), 106–118.
16. Carlsson, A. E. (2010). Actin dynamics: from nanoscale to microscale. Annual Review of Biophysics, 39, 91–110.
17. Carrasquilla, J., & Melko, R. G. (2017). Machine learning phases of matter. Nature Physics, 13(5), 431–434.
18. Castellana, M., Wilson, M. Z., Xu, Y., Joshi, P., Cristea, I. M., Rabinowitz, J. D., ...&Wingreen, N. S. (2014). Enzyme clustering accelerates processing of intermediates through metabolic channeling. Nature Biotechnology, 32(10), 1011–1018.
19. Castellano, C., Fortunato, S., & Loreto, V. (2009). Statistical physics of social dynamics. Reviews of Modern Physics, 81(2), 591.
20. Chan, V., Asada, H. H., & Bashir, R. (2014). Utilization and control of bioactuators across multiple length scales. Lab on a Chip, 14(4), 653–670.
21. Chandler, D. (1987). Introduction to modern statistical. Mechanics. Oxford University Press (Vol.1, pp. 1–33), Oxford, UK.
22. Ciarlet, P. G. (2005). An introduction to differential geometry with applications to elasticity. Journal of Elasticity, 78(1–3), 1–215.
23. Cocco, S., Feinauer, C., Figliuzzi, M., Monasson, R., & Weigt, M. (2018). Inverse statistical physics of protein sequences: a key issues review. Reports on Progress in Physics, 81(3), 032601.
24. Collin, D., Ritort, F., Jarzynski, C., Smith, S. B., Tinoco, I., & Bustamante, C. (2005). Verification of the Crooks fluctuation theorem and recovery of RNA folding free energies. Nature, 437(7056), 231–234.
25. Cover, T. M., & Thomas, J. A. (2012). Elements of Information Theory, Vol.1, pp. 1–43.

26. Crooks, G. E. (1999a). Entropy production fluctuation theorem and the nonequilibrium work relation for free energy differences. Physical Review E, 60(3), 2721.
27. Crooks, G. E. (2009b). Comment regarding "On the Crooks fluctuation theorem and the Jarzynski equality." J. Chem. Phys. 129, 091101 (2008); and "Nonequilibrium fluctuation-dissipation theorem of Brownian dynamics." J. Chem. Phys. 129, 144113 (2008). The Journal of chemical physics, 130(10), 03B801.
28. Deffner, S., & Lutz, E. (2012). Information free energy for nonequilibrium states, 1, 1–30. arXiv preprint arXiv:1201.3888.
29. England, J. L. (2013). Statistical physics of self-replication. The Journal of Chemical Physics, 139(12), 09B623_1.
30. Feng, G., Cheng, Y., Wang, S. Y., Hsu, L. C., Feliz, Y., Borca-Tasciuc, D. A., ... & Moraru, C. I. (2014). Alumina surfaces with nanoscale topography reduce attachment and biofilm formation by Escherichia coli and Listeria spp. Biofouling, 30(10), 1253–1268.
31. French, J. B., Jones, S. A., Deng, H., Pedley, A. M., Kim, D., Chan, C. Y., ... & Huang, T. J. (2016). Spatial colocalization and functional link of purinosomes with mitochondria. Science, 351(6274), 733–737.
32. Frey, E. (2013). Experimental Free Energy Landscape Reconstruction of DNA Unstacking Using Crooks Fluctuation Theorem (Vol.1, pp. 1–32).
33. Friedman, J., Hastie, T., & Tibshirani, R. (2001). The Elements of Statistical Learning (Vol. 1, No.10, pp. 1–40). New York: Springer series in statistics.
34. Friston, K. (2012). The history of the future of the Bayesian brain. NeuroImage, 62(2), 1230–1233.
35. Fu, R., Sutcliffe, D., Zhao, H., Huang, X., Schretlen, D. J., Benkovic, S., & Jinnah, H. A. (2015). Clinical severity in Lesch–Nyhan disease: the role of residual enzyme and compensatory pathways. Molecular Genetics and Metabolism, 114(1), 55–61.
36. García-Palacios, J. L., & Lázaro, F. J. (1998). Langevin-dynamics study of the dynamical properties of small magnetic particles. Physical Review B, 58(22), 14937.
37. Gelman, A., Carlin, J. B., Stern, H. S., Dunson, D. B., Vehtari, A., & Rubin, D. B. (2014). Bayesian Data Analysis, Vol. 2, pp. 1–30. CRC Press. Boca Raton, FL.

38. Goldt, S., & Seifert, U. (2017). Thermodynamic efficiency of learning a rule in neural networks. New Journal of Physics, 19(11), 113001.
39. Gomez-Marin, A., Schmiedl, T., & Seifert, U. (2008). Optimal protocols for minimal work processes in underdamped stochastic thermodynamics. The Journal of Chemical Physics, 129(2), 024114.
40. Gorban, A. N., & Tyukin, I. Y. (2018). Blessing of dimensionality: mathematical foundations of the statistical physics of data. Philosophical Transactions of the Royal Society A: Mathematical, Physical and Engineering Sciences, 376(2118), 20170237.
41. Ito, S., & Sagawa, T. (2016). Information flow and entropy production on Bayesian networks. Mathematical Foundations and Applications of Graph Entropy, 3(2), 1–30.
42. Jarzynski, C. (1997). Nonequilibrium equality for free energy differences. Physical Review Letters, 78(14), 2690.
43. Jørgensen, K., Rasmussen, A. V., Morant, M., Nielsen, A. H., Bjarnholt, N., Zagrobelny, M., ...&Møller, B. L. (2005). Metabolon formation and metabolic channeling in the biosynthesis of plant natural products. Current Opinion in Plant Biology, 8(3), 280–291.
44. Kecman, V. (2001). Learning and Soft Computing: Support Vector Machines, Neural Networks, and Fuzzy Logic Models (Vol.1, pp. 1–33). MI Press.
45. Koller, D., & Friedman, N. (2009). Probabilistic graphical models: principles and techniques (Vol. 1, pp. 1–30). MIT Press.
46. Krem, M. M., & Di Cera, E. (2002). Evolution of enzyme cascades from embryonic development to blood coagulation. Trends in Biochemical Sciences, 27(2), 67–74.
47. Lavendhomme, R. (2013). Basic Concepts of Synthetic Differential Geometry (Vol. 13, pp. 1–30). Springer Science & Business Media.
48. Leimkuhler, B., Matthews, C., & Vlaar, T. (2019). Partitioned integrators for thermodynamic parameterization of neural networks (Vol.1, pp. 1–22). arXiv preprint arXiv:1908.11843.
49. Linsker, R. (1988). Self-organization in a perceptual network. Computer, 21(3), 105–117.
50. Lu, Z., Mandal, D., & Jarzynski, C. (2014). Engineering Maxwell's demon. Physics Today, 67(8), 60–61.

51. Malzahn, D., & Opper, M. (2005). A statistical physics approach for the analysis of machine learning algorithms on real data. Journal of Statistical Mechanics: Theory and Experiment, 2005(11), P11001.
52. Manasse, F. K., & Misner, C. W. (1963). Fermi normal coordinates and some basic concepts in differential geometry. Journal of Mathematical Physics, 4(6), 735–745.
53. McHale, J. M., Auroux, A., Perrotta, A. J., & Navrotsky, A. (1997). Surface energies and thermodynamic phase stability in nanocrystallinealuminas. Science, 277(5327), 788–791.
54. Mestres, P., Martinez, I. A., Ortiz-Ambriz, A., Rica, R. A., & Roldan, E. (2014). Realization of nonequilibrium thermodynamic processes using external colored noise. Physical Review E, 90(3), 032116.
55. Metropolis, N., Rosenbluth, A. W., Rosenbluth, M. N., Teller, A. H., & Teller, E. (1953). Equation of state calculations by fast computing machines. The Journal of Chemical Physics, 21(6), 1087–1092.
56. Mezard, M., & Montanari, A. (2009). Information, Physics, and Computation (Vol.1, pp. 1–22). Oxford University Press.
57. Mézard, M., Parisi, G., & Virasoro, M. (1987). Spin Glass Theory and Beyond: An Introduction to the Replica Method and Its Applications (Vol. 9, pp. 1–30). World Scientific Publishing Company.
58. Moore, C. (2017). The computer science and physics of community detection: Landscapes, phase transitions, and hardness, 1, 1–33. arXiv preprint arXiv:1702.00467.
59. Muddana, H. S., Sengupta, S., Mallouk, T. E., Sen, A., & Butler, P. J. (2010). Substrate catalysis enhances single-enzyme diffusion. Journal of the American Chemical Society, 132(7), 2110–2111.
60. Neal, R. M. (2011). MCMC using Hamiltonian dynamics. Handbook of markov chain monte carlo, 2(11), 2–12.
61. Novikov, S. P., & Fomenko, A. T. (2013). Basic Elements of Differential Geometry and Topology (Vol. 60, pp. 1–30). Springer Science & Business Media.
62. Paduano, L., Sartorio, R., & Vitagliano, V. (1998). Diffusion coefficients of the ternary system α-cyclodextrin– sodium benzenesulfonate– water at 25 C: the effect of chemical equilibrium and complex formation on the diffusion coefficients of a ternary system. The Journal of Physical Chemistry B, 102(25), 5023–5028.

63. Parrondo, J. M., Horowitz, J. M., & Sagawa, T. (2015). Thermodynamics of information. Nature physics, 11(2), 131–139.
64. Postma, M., & Van Haastert, P. J. (2001). A diffusion–translocation model for gradient sensing by chemotactic cells. Biophysical Journal, 81(3), 1314–1323.
65. Reguera, D., Rubi, J. M., & Vilar, J. M. G. (2005). The Mesoscopic Dynamics of Thermodynamic Systems, 1, 1–30.
66. Riedel, C., Gabizon, R., Wilson, C. A., Hamadani, K., Tsekouras, K., Marqusee, S., ...& Bustamante, C. (2015). The heat released during catalytic turnover enhances the diffusion of an enzyme. Nature, 517(7533), 227–230.
67. Robbins, H., & Monro, S. (1951). A stochastic approximation method. The Annals of Mathematical Statistics, 1, 400–407.
68. Roudi, Y., Aurell, E., & Hertz, J. A. (2009). Statistical physics of pairwise probability models. Frontiers in Computational Neuroscience, 3, 22.
69. Sagawa, T., & Ueda, M. (2010). Generalized Jarzynski equality under nonequilibrium feedback control. Physical Review Letters, 104(9), 090602.
70. Seifert, U. (2012). Stochastic thermodynamics, fluctuation theorems and molecular machines. Reports on Progress in Physics, 75(12), 126001.
71. Sengupta, S., Dey, K. K., Muddana, H. S., Tabouillot, T., Ibele, M. E., Butler, P. J., & Sen, A. (2013). Enzyme molecules as nanomotors. Journal of the American Chemical Society, 135(4), 1406–1414.
72. Soliman, S., Pagliari, S., Rinaldi, A., Forte, G., Fiaccavento, R., Pagliari, F., ...&Traversa, E. (2010). Multiscale three-dimensional scaffolds for soft tissue engineering via multimodal electrospinning. Actabiomaterialia, 6(4), 1227–1237.
73. Van den Broeck, C. (2013). Stochastic thermodynamics: A brief introduction. Physics of Complex Colloids, 184, 155–193.
74. Van Der Aalst, W. (2016). Data science in action. In Process Mining (Vol.1, pp. 3–23). Springer, Berlin, Heidelberg.
75. Vanag, V. K., & Epstein, I. R. (2009). Cross-diffusion and pattern formation in reaction–diffusion systems. Physical Chemistry Chemical Physics, 11(6), 897–912.

76. Vapnik, V. N. (1999). An overview of statistical learning theory. IEEE Transactions on Neural Networks, 10(5), 988–999.
77. Vergara, A., Paduano, L., & Sartorio, R. (2002). Mechanism of Protein– Poly (ethylene glycol) Interaction from a Diffusive Point of View. Macromolecules, 35(4), 1389–1398.
78. Vlcek, L., Ziatdinov, M., Maksov, A., Tselev, A., Baddorf, A. P., Kalinin, S. V., & Vasudevan, R. K. (2019). Learning from imperfections: predicting structure and thermodynamics from atomic imaging of fluctuations. ACS Nano, 13(1), 718–727.
79. Wasserman, L. (2013). All of Statistics: A Concise Course in Statistical Inference (Vol.1, pp. 1–33). Springer Science & Business Media.
80. Watkin, T. L., Rau, A., & Biehl, M. (1993). The statistical mechanics of learning a rule. Reviews of Modern Physics, 65(2), 499.
81. Welling, M., & Teh, Y. W. (2011). Bayesian learning via stochastic gradient Langevin dynamics. In Proceedings of the 28th International Conference on Machine Learning (ICML-11) (Vol.1, pp. 681–688).
82. Wolpert, D. H. (2006). Information theory—the bridge connecting bounded rational game theory and statistical physics. In Complex Engineered Systems (Vol. 1, pp. 262–290). Springer, Berlin, Heidelberg.
83. Wu, F., & Minteer, S. (2015). Krebs cycle metabolon: structural evidence of substrate channeling revealed by cross-linking and mass spectrometry. AngewandteChemie International Edition, 54(6), 1851–1854.
84. Yuille, A. L. (1990). Generalized deformable models, statistical physics, and matching problems. Neural Computation, 2(1), 1–24.
85. Yuille, A. L., Stolorz, P., & Utans, J. (1994). Statistical physics, mixtures of distributions, and the EM algorithm. Neural Computation, 6(2), 334–340.
86. Zdeborová, L., & Krzakala, F. (2016). Statistical physics of inference: Thresholds and algorithms. Advances in Physics, 65(5), 453–552.
87. Zhao, H., French, J. B., Fang, Y., & Benkovic, S. J. (2013). The purinosome, a multi-protein complex involved in the de novo biosynthesis of purines in humans. Chemical communications, 49(40), 4444–4452.
88. Zhao, X., Palacci, H., Yadav, V., Spiering, M. M., Gilson, M. K., Butler, P. J., ...& Sen, A. (2018). Substrate-driven chemotactic assembly in an enzyme cascade. Nature Chemistry, 10(3), 311.

Chapter 4

Particle Physics and Cosmology

CONTENTS

4.1. INTRODUCTION

A diversified portfolio of progressing and on schedule experiments is self-possessed and well-composed to explore the wide universe from the tiniest of the unbelievable world of particles to the wonderments of the widespread universe. To compare the predictions, calculated by specific theoretical models, immense data is provided by the (LSST) Large Synoptic Survey Telescope and the (LHC) Large Hadron Collider. Some entrenched physical models are covering both areas, these may play a part as null hypotheses: a conventional model of ΛCDM cosmology and particle physics, which includes a cosmological constant Λ and cold dark matter. Interestingly, hypotheses such as general relativity and quantum field theory are regarded as the alternate hypothesis but their formulation is based on the same theoretical frameworks. Despite such precise theoretical tools, from the null, the expected deviations are intimidating and are inconceivably small and they are revealing such ultra-fine effects that may need a vigorous treatment of experimental apparatuses. The statistical inference is complicated as for the data, the high-fidelity predictions do come in the complex computer simulations but not in the simple closed-form equations.

Figure 4.1. The contribution of machine learning in particle physics and cosmology.

[Source: https://www.symmetrymagazine.org/article/studying-the-stars-with-machine-learning.]

In the branch of cosmology and particle physics, machine learning is generating a significant impact. As machine learning offers new bold techniques that are capable of not only confronting these challenges but also motivates to introduce strategies with a newer perspective which can serve as a solution. These changes significantly span both the aspects of these fields, the experimental and theoretical. Moreover, long term transformational changes and the prospect of immediate impact are also covered by these changes.

4.2. THE SIMULATION'S ROLE

In the field of cosmology and particle physics, machine learning has an important aspect as labeled training data $\{X_{\downarrow}\mu, Y_{\downarrow}\mu\}_{\downarrow}(\mu = 1)^{\uparrow} n$ samples can be generated by using computer simulations. For example, the time when any target for instance y referring to a parameter that may appear in the fundamental theory, particular scattering process, or a particle type, can be directly specified in the simulation code. This helps in the direct sampling of the simulation X ~p (.|y). Contrary to it, in other cases, samples are provided (X, Z) ~ p (.), but these provided samples' simulation does not directly depend on y. In these cases, Z serves as latent variables and is used to describe the inside happening of the simulation that cannot be observed in an actual experiment. From the simulation, we can create labeled training data $\{X_{\downarrow}\mu, [(Y(Z))]_{\downarrow}\mu\}_{\downarrow}(\mu = 1)^{\uparrow} n$ but it is possible only when the latent variables can be used to compute the target label through a function y (Z). For generating labeled training data, high- fidelity simulation usage has served as a key element in the initial success of supervised learning, not only in these areas, it has also focused on the research that discusses the limitations of this approach.

A suite based on high-fidelity simulations has been developed by particle physicists; they have composed the simulations hierarchically so that their interaction can be described over a wide range of length scales.

In these simulations, components like the phenomenological model for radiation's complex patterns, expansion of Feynman's diagrammatic perturbative quantum field theory, and other models for matter and particle interaction (in the detector) are included. The simulations, as a result, have high fidelity. There are free parameters in the simulation, these parameters can be tuned. These simulations also have numerous residual uncertainties that must be reckoned while performing down-stream analysis tasks.

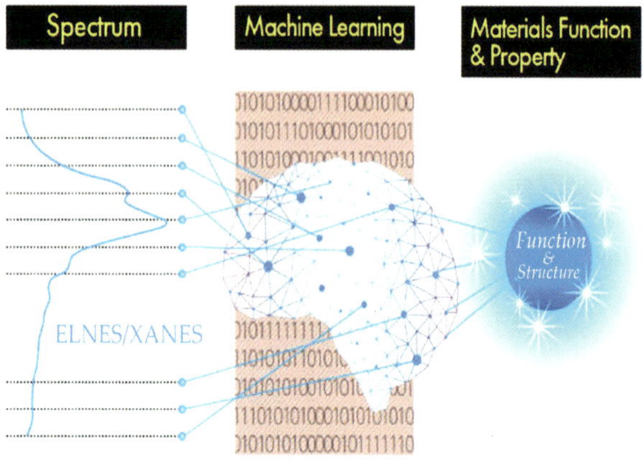

Figure 4.2. Working arrangement of ML in particle Physics.

[Source: https://phys.org/news/2019-03-machine-reveals-rapid-material-classification.html.]

Similarly, the universe's evolution can be simulated by cosmologists. By using matter's relevant non-gravitational effects and general relativity and also radiation, which serves as an important factor during structure formation, cosmologists can simulate it at different scales. In specific settings, an abundant array of approximations is present. These arrays offer enormous speedups, in contrast to the massive object's N -body simulations, it is computationally expensive to calculate billions of objects interacting gravitationally, especially when effects of non-gravitational feedback are included, it makes it prohibitively expensive.

Because of their expensiveness, N -body simulations are relatively few. A very large space-time volume is covered by these few simulations. On a large scale, this volume is statistically homogeneous and isotropic. In comparison, simulations for particle physics are stochastic throughout irrespective of phase, initially from high-energy scattering to the final stage when in a detector, there are low-energy interactions, simulations remain stochastic. In high-energy collider experiments, simulations can run in a parallel manner on commodity hardware but a huge quantity of simulated collisions are required by the physics goals.

In these fields, as simulations have a critical role, mostly the new researches in the field of machine learning are interlinked to simulation. The major goals of these researches are to:

- To develop more data-efficient techniques by integrating the machine learning models and the basic domain knowledge of this field;
- To include the uncertainties present in the simulations into the procedure of training;
- Instead of relying on the simulation, develop procedures that can be weakly supervised and practiced on real data;
- For finding anomalous features, develop algorithms that can be used for anomaly detection, without simulating a specific signal hypothesis;
- To improve the simulation's tuning, the simulated data should be adjusted or reweighted so that it matches the real data, or for modeling residuals between the real data and the simulation, use machine learning;
- For generating synthetic data quickly, learn fast neural network can serve as a replacement for simulation;
- Develop techniques such as approximate inference techniques as they make the use of simulation efficient; and
- Learn a fast neural network that can serve as a replacement for statistical inference.

4.3. REGRESSION AND CLASSIFICATION IN PARTICLE PHYSICS

In experimental particle physics, techniques of machine learning are in use from decades, these techniques are used to aid event selection and particle identification, these tasks fall under the category of classification. The machine learning technique is also used for reconstruction, this task falls under the category of regression. For training a predictive model, supervised learning is used, these models are based on a huge number of training samples that are labeled , $\{X_\downarrow \mu, Y_\downarrow \mu\}_\downarrow (\mu = 1)^\uparrow n$ where input data is denoted by X, and the target label is denoted by y. When dealing with particle identification, localized energy accumulations (found in the detector) are characterized by the input features X and y denote any of particle species such as a photon, pion, electron, etc. Almost the same kind of sensor data X is utilized in the reconstruction task, but in these tasks, the target label y denotes to the momentum or energy of the particle which is responsible behind all those energy deposits. We can apply these algorithms to bulk data processing (LHC data). While selecting an event, a subset of most relevant collisions for a specific analysis task is targeted and this task is termed as event selection. For instance, while searching for the dark matter, Higgs boson, and supersymmetry, a subset of the LHC data must be selected by data analysts, this subset is consistent with the attributes of hypothetical "signal" processes. Usually, so-called "background" processes mimicking the attributes of the signal either because of quantum mechanical effects or experimental limitations, satisfy the event selection requirements. Simplest forms of searches are reduced to contrast with various events in the data. The requirements are satisfied by these searches to the predictions of signal-plus-background alternate and a background- only null hypothesis. Hence, the more effective and efficient the event selection requirements are at accepting signal processes and rejecting background processes, the stronger will be the resulting statistical analysis. The classification techniques of machine learning, within high-energy physics, are conventionally termed as multivariate analysis for emphasizing the comparison of traditional techniques that are formed by depending on simple thresholding applied on engineered or carefully selected features.

For these tasks, in the time frame of the 1990s and initial phase of the 2000s, the commonly used techniques were simple feed-forward neural networks. Boosted Decision Trees (BDTs) displaced the neural networks, from more than a decade, as a substitute for regression and classification tasks (Breiman et al., 1984; Freund and Schapire, 1997; Roe et al., 2005). Around 2014, deep learning techniques emerged and were proven more powerful and efficient in many applications (Guest et al., 2018; Radovic et al., 2018).

Initially, deep learning was used for the task of event-selection, hypothesized particles were targeted from the theories. The result was splendid as it outmatched boosted decision trees and even engineered features for achieving this great performance were not required (Baldi et al., 2014). In this viability study work, a deep multi-layer perceptron network resulted, and it was trained by using a large-scale training set, under a simple detector setup. Soon after this, the concept of parameterized classifier evolved, this concept extended the idea of a binary classifier to a situation where signal hypothesis in which, y =1 is lifted towards a composite hypothesis in which continuous parameterization is observed, for example, hypothesized particle's mass (Baldi et al., 2016b).

4.3.1. Jet Physics

In hadron collisions, for instance like the LHC, are produced high energy gluons and quarks. These gluons and quarks radiate more gluons and quarks that in due course combine into neutral-color composite particles, they combine because of confinement phenomena. The ensuing collimated "spray" of baryons and mesons, striking the detector is referred to as a jet. Developing theoretically robust characterization of a jet's structure that can serve in testing the predictions coming from quantum chromodynamics (QCD), as for decades, it has been a popular area of research in particle physics. Moreover, the particle production decaying in more than one jet for physics is predicted by numerous scenarios Beyond the Standard Model. If the produced particles (unstable ones) have large momentum, in this scenario the jets (resulting ones) are boosted, they are boosted to a level that they overlap and combine to form a single fat jet that has a nontrivial substructure. These fat jets or boosted jets can be classified from the more abundantly produced jets by the processes involving gluons and quarks, it is an area capable of significantly improving the physics of LHC. Jet tagging is a classification task that identifies the progenitor for a jet.

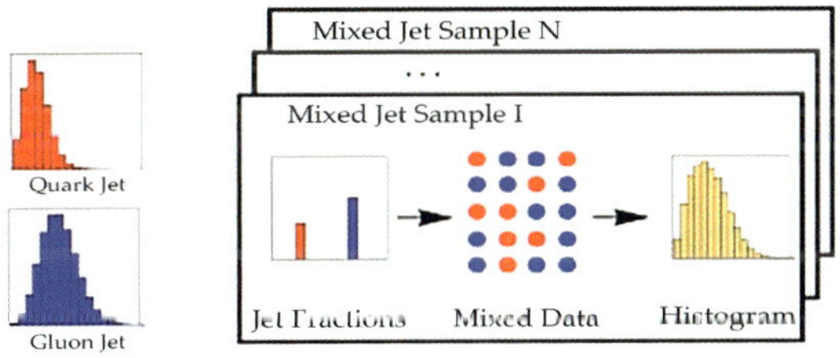

Figure 4.3. Machine learning applications in Jet-Physics.

[Source: https://www.ericmetodiev.com/files/slides/Harvard_MLJets_EMM.pdf.]

To perform jet tagging, soon after the initial use of applications for event selection based on deep learning, applications based on deep convolutional networks were adopted for this purpose, in them, the detector data that is low-level lends itself to a representation that is like an image (Baldi et al., 2016a; de Oliveira et al., 2016). Within particle physics, machine learning techniques are in use for decades, but this practice is restricted to input features X, as it has a fixed dimensionality.

In jet physics, one of the challenges is that the data's natural representation is in particle terms, and their association with particle varies as the different number of particles can be associated with a jet. In particle physics, the recurrent neural network's first application was regarding flavor tagging (Guest et al., 2016). Recently the researches have shifted more towards the usage of different network architectures, among them, recurrent networks operating on graphs, trees, and sequences are included (Louppe et al., 2016; Larkoski et al., 2020). Hybrid approaches are included in them, which, in designing the architecture, leverages domain knowledge. For instance, inspired by the natural language processing techniques, recursive networks were created and designed for operating over tree-structures originated from jet clustering algorithms (Louppe et al., 2019a; 2019b). Similarly, invariance has been a major motivation behind many networks that stabilizes the particle's radiation pattern and its details, and the particles are permutated when presented to the network (Komiske et al., 2018a; 2018b, 2019). Recently, for specific benchmark problems, different approaches have been compared and organized (Kasieczka et al., 2019). Besides regression and classification, techniques of machine learning have been used to model smooth spectra and for density estimation. The spectra are modeled when there is no well-motivated analytical form and when significant uncertainties are present in the simulation (Frate et al., 2017). Instead of using any specific concrete physical model, this allows us to use a diffuse prior for modeling of alternative signal hypotheses. More abstractly, in this work, the Gaussian process is being used for the intensity modeling of inhomogeneous Poisson point process; this scenario is usually found in cosmology, astrophysics, and particle physics. Another interesting thing about this work is that using compositional rules, we can construct the Gaussian process kernels. These compositional rules are in correspondence with the causal model physicists instinctively utilized for describing the observation aiding in interpretability (Duvenaud et al., 2013).

4.3.2. Neutrino Physics

The interaction of matter and neutrinos is very feeble, thus large detector volumes are required by the experiments for achieving appreciable interaction rates. In the detector volume, various patterns of energy deposits (localized) are left by the different interactions irrespective from where they are coming, whether from background cosmic ray processes or distinct species of neutrinos. The volume of the detector is homogeneous, in this way usage of convolutional neural networks is motivated.

The first deep convolutional network application in the data analysis was in the perspective of the NOvA experiment: scintillating mineral oil was used in it. Imaging from two vantage points, NOvA interaction leads to light production. A convolutional network was developed by NOvA that processes two images simultaneously (Aurisano et al., 2016). For the same purity, the efficiency of selecting electron neutrinos is improved by 40% by their network. This network is used in searches related to the hypothetical sterile neutrino and for electron neutrino's appearance.

Similarly, neutrinos created at Fermilab are detected by the MicroBooNE experiment. A time projection chamber of liquid-argon of 170 tons is used. The liquid argon is ionized by the charged particles and drift of the ionization electrons to three wire planes is observed. The resulting data undergoes processing and is then represented by an image of 33-megapixel that is chiefly populated by noise and sparsely populated by deposits of legitimate energy. A Faster RCNN was used by the MicroBooNE collaboration (Ren et al., 2015) for identifying and for localizing neutrino interactions by using bounding boxes (Acciarri et al., 2017). This approach is crucial for the future, as it will help the neutrino experiments based on the time projection chamber of liquid-argon, e.g., DUNE, Deep Underground Neutrino Experiment.

We are well aware of the use of machine learning in producing low energy neutrinos at accelerator facilities, besides it, machine learning is also used for studying high energy neutrinos by the use of the Ice- Cube observatory (situated in the South Pole). In specific, graph neural networks and 3D convolutional are employed to problems related to signal classification. In the hindmost approach, edges serve as sensors' spatial coordinate's learned function and vertices serve as sensors to model the detector array as a graph. 3D convolutional neural network and a traditionally used physics-based method, both were outperformed by the graph neural network (Choma et al., 2018).

4.3.3. Robustness to Systematic Uncertainties

Physicists working on experimental particles well know the fact that regardless of how incredibly accurate a simulation is, it is never perfect. Consequently, falling into two major classes, numerous strategies have been developed by the community. The first one incorporates the mis-modeling effect when we use the simulation for training. The analysis chain and propagating the underlying uncertainty sources (for example, the proton's gluon and quark composition, calibrations, higher-order correction's impact from the theory of perturbation, and detector response) through the simulation, both of them are involved. For each of these detector response sources, a nuisance parameter, represented as ν is included, and these nuisance parameters, parameterize the resulting statistical model p(X y, ν). Additionally, the data's probability function is augmented with p (ν) which is a term that represents the uncertainty in the sources. In the machine learning context, regressors and classifiers are usually trained by utilizing the nominal simulation generated data $\nu = \nu_0$: a predictive model f (X ν_0) is yielded. If we treat this model as fixed, the uncertainty in ν can be propagated through f (X ν_0) with the help of the model p(X y, ν) p (ν). Nonetheless, based on this approach, the down-stream analysis (statistical) is not optimal, this is due to the reason that while training the model, the uncertainty on ν was not taken into account.

In machine learning terms, this situation is called a covariate shift among domains and it is represented by the target distribution ν and the training distribution ν_0. Different techniques for domain adaptation are available for classifiers training but are restricted, to train the target of binary domains ν. An adversarial training technique for addressing this problem was developed, in which the domain adaptation is extended to ν ε ℝq parametrized domains (Louppe et al., 2016). The network is encouraged by the adversarial approach for learning a pivotal quantity, in it, p(f (X)| y, ν) does not depend on ν, or equivalently p(f (X), ν| y) = p(f (X)| y)p(ν). This approach is also used in the algorithmic fairness context. In this approach, one desires for training a regressor or classifiers independently, without using observable quantities or specific continuous attributes. For example, in jet physics, a jet tagger working from the jet invariant mass will be more preferred (Shimmin et al., 2017). In the past, for boosted decision trees, an algorithm named uboost was developed and used for achieving similar goals (Rogozhnikov et al., 2015; Stevens and Williams, 2013).

Generally, within particle physics, another strategy commonly used in coping up with systematic mis-modeling is to avoid the simulation's usage while distribution modeling p(X| y). An index over numerous data subsets is denoted by R, it includes the data that is satisfying corresponding selection requirements. In control regions, numerous data-driven strategies are developed for relating the distributions of the data, p(X| y, R = 0), to p(X| y, R = 1), which shows the distributions present in the regions of interest. The simulation is also involved in these relationships, but this approach aims to base relationships on the simulation's aspect (the ones considered robust are prioritized). For instance, for any specific process called y, the distribution estimation p(X| y, R = 1) through identifying a data subset R = 0 and is dominated by p(y| R = 0) 1 and y. This extreme situation is limited in terms of applicability. Lately, weakly supervised techniques are developed. They identify the regions in which the class proportions are well known or they assume that the p(y| R) (the relative probabilities) are linearly independent (Komiske et al., 2018a; Metodiev et al., 2017). It is also assumed by the techniques that the p(X| y, R), distributions do not depend on R, which is questionable in some contexts and reasonable in some. This approach is also used for training the jet taggers, discriminating between gluons and quarks; it is an area where the simulation's fidelity does not remain adequate. For particle physics, this approach is an important and vital development for machine learning, though few problems are present in it. For instance, we cannot apply this approach if the y (target categories) is in correspondence of a hypothetical particle that is missing from data or does not exist.

4.3.4. Triggering

LHC like collider experiments collects huge amounts of data, as they target the exceedingly rare phenomena. The phenomena that are already characterized and studied, are involved by the bulk of the collisions, and the full data stream that associates the data volume to be impractically large. Consequently, a real-time system for data- reduction, terms a trigger is used by the collider experiments. The critical decisions about discarding and keeping the events are taken by the triggers. For every 100,000 events, only about 1 event is retained by the CMS and AT- LAS experiments. In these systems, the techniques of machine learning are used to different degrees. Essentially, in this context, the same particle classification (identification) tasks appear, though, in the real-time environment, the performance and computational demands in terms of false negatives and positives are different.

In the trigger, the LHCb experiment has served as a leader in using techniques of machine learning. Machine learning algorithms select around 70% of the total selected data by the LHC trigger. Initially, for this purpose, a boosted decision tree was used in the experiment (Gligorov and Williams, 2013), later, Yandex developed the MatrixNet algorithm that replaced it (Likhomanenko et al., 2015).

Specialized firmware and hardware are often used by the Trigger systems, for instance, FPGAs (field-programmable gate arrays). In recent times, tools for streamlining the compilation of models based on machine learning are developed for FPGAs for addressing real-time triggering system's requirements (Tsaris et al., 2018).

4.3.5. Theoretical Particle Physics

Although most of the machine learning that is applied in cosmology and particle physics is focused on analyzing the observational data, it is not the only use case for machine learning. Machine learning is also used in theoretical physics. For example, machine learning was used as a tool in condensed matter or many-body quantum physics. Also, machine learning was used for characterizing the landscape of string theories (Carifio et al., 2017) and to understand the AdS/CFT similarity (Hashimoto et al., 2018). In the perspective of lattice quantum chromodynamic, deep learning was also used. A more specific example would be: For predicting the parameters in QCD Lagrangian from lattice configurations, deep neural networks were used. For improving the efficiency of LQCD calculations (that are computationally intensive), there was a need for multi-scale action machine learning approaches to handle the complex calculation. The problem was represented as a regression task, the greatest challenge that the author faced was that it had relatively few training instances. A possible way to solve the task with few training examples was to make use of local gauge symmetries present in the data and to have a strong grip over known space-time. Since there was much richness in the symmetries, data augmentation could not be used. Therefore, the author decided to feature engineer resulting in gauge symmetry and translational invariance of space-time. Although this approach was a success, however, it is highly recommended to make use of different classes of networks that are covariant (or equivariant) to the symmetry (these approaches are discussed in more detail in Sec. III.F). Argon Leadership Computing Facility is supporting the continuation of this work, their new Intel-Cray system Aurora will have the computational power to perform over 1 exaflop and this system is targeting challenges that combine modern machine learning techniques with traditional high-performance computing.

4.4. REGRESSION AND CLASSIFICATION IN COSMOLOGY

4.4.1. Photometric Redshift

Overtime universe expansion leads to the red-shifting of distant luminous objects, and in observational cosmology, this relation of the distance-redshift is a major component. We can get an accurate estimation of redshift through spectroscopy, but this methodology is time-consuming and is expensive. A rough estimation of the spectral energy distribution is given by photometric surveys based on imaging in color bands or broadband photometry. The regression task of redshifts estimation from photo-metric data is called photometric redshift. In such a case, ground truth training instances are gathered from accurate spectroscopic surveys.

Photometric redshift traditional approaches are based on methods of template fitting (Benítez, 2000; Brammer et al., 2008; Feldmann et al., 2006). For over a decade, machine learning methods based on boosted decision trees and neural networks, have been used by cosmologists for photometric redshift (Carrasco Kind and Brunner, 2013; Collister and Lahav, 2004; Firth et al., 2003). For the redshift, a compelling aspect of their work is the attempt that was made which resulted in going beyond a point estimate. Multiple approaches already exist for determining the unpredictability of this estimate (redshift) and to acquire a posterior distribution.

Although the training instances are not created from a simulation, a concern exists that the training set distribution may differ from the representative of the distribution of instances that will be applied to the learning models. This might be because of different selection, as it affects the spectroscopic survey and understatement in the photometric surveys. Various such methodologies were considered by the Dark Energy Survey resulting in the establishment of a validation process to assess them critically (Bonnett et al., 2016). Work has been done to be robust to these differences by making use of hierarchical models to build in additional causal structure in the models. In terms of machine learning, these newly constructed models help in domain adaptation and transfer learning. These models also aim to merge the accountability of template fitting approaches and the flexibility of models based on machine learning (Leistedt et al., 2018).

For this supervised classification problem, methods of basic feedforward, residual neural networks, and convolutional neural networks have been applied (Estrada et al., 2007; Lanusse et al., 2018; Marshall et al., 2009). In this experiment, the training instances were collected from simulation using Pipeline for Images of Cosmological Strong (PICS), LensPop (Collett, 2015) for mock LSST observing, and lensing (Li et al., 2016) for the lensing ray-tracing. Lensing object characterization by using maximum probability estimation is a non-linear optimization and computationally intensive task, once it is spotted. For quick parameter estimation of the Singular Isothermal

Ellipsoid density profile, which is commonly used for modeling the strong lensing systems, convolutional networks have been recently used (Hezaveh et al., 2017).

4.4.2. Gravitational Parameter Estimation and Lens Finding

Gravitational lensing, in which a huge forefront object covers the background object's image, is one of the most noticeable effects of general relativity. Powerful gravitational lensing takes place, for instance, in the sky, when a huge forefront galaxy is almost coincident with a background source. These occurrences are a strong probe of the distribution of dark matter of huge galaxies and can give valuable cosmological constraints. Due to the rarity of such systems, a reliable and expandable lens finding system is much needed so that it can cope up with surveys such as WFIRST, LSST, and Euclid. To improve the situation, easy to acquire huge labeled training instances set, data augmentation methodologies, and different types of subsampling have been explored. The backdrop is also a considerable alternative way of subsampling, the loss function's stochastic gradients are provided by it, and it also makes it available for individual instances by establishing stochastic masking in the pipeline of backpropagation (Golkar and Cranmer, 2018).

In a classification setting, inference on the foundational cosmological model can also emerge. Specifically, for weak lensing observable prediction, predicted by the ΛCDM model can be imitated by modified gravity models with huge neutrinos. By including higher-order statistics or any weak lensing signal representations, we can break the degeneracies that are faced when limiting the Xμ to second-order statistics. Specifically, the writer of (Peel et al., 2018) created a novel description of the weak lensing signal 's decomposing wavelet as an input to a convolutional network. The resulted methodology was able to differentiate between earlier degenerate models with an accuracy of 83%–100%.

To calculate the mass of galaxy clusters deep learning has also been used for this task, in the universe, galaxy clusters are the greatest gravitationally bound structures and a strong cosmological probe. Because most of these galaxy clusters appear in the form of dark matter it makes it difficult to directly observe them. Masses of galaxy clusters can be calculated through dynamical analysis of the galaxies of a cluster, gravitational lensing, or the intracluster medium's X-ray observations. Support Distribution Machines (Póczos et al., 2012) was the first use of machine learning for applying dynamical clustering mass estimation on a dark-matter simulation

(Ntampaka et al., 2015, 2016). By making use of MACSIS (Henson et al., 2016) simulation multiple non-neural network algorithms such as kernel ridge regression (Gaussian process regression), gradient boosted tree regressors, support vector machines and others have been made use of in this problem for training instances. This simulation is not only a dark- matter-only simulation but it also includes the impact of different astrophysical processes and permits the development of a real-world processing pipeline that can be used on observational data. Huge surveys such as DESI, eROSITA, eBOSS, SPT-3G, Euclid, and Act- Pol motivated the need for an automated and accurate mass estimation pipeline. The writer deduced that by making use of machine learning techniques, the predicted-to-true mass ratio is reduced by almost a factor of 4, compared to the traditional σ M relation (Armitage et al., 2019). In recent times, convolutional neural networks have been utilized in the virial scaling relation for alleviating systematics, further enhancing dynamical mass estimates (Ho et al., 2019). Other networks such as convolutional neural networks are utilized to calculate cluster masses using synthetic observations of X-ray of galaxy clusters, this resulted in the observation that by comparison with the X-ray luminosity based methods, in the predicted mass, the scatter is reduced (Ntampaka et al., 2018).

4.4.3. Other Examples

With regards to the previous examples, wherewith a more time-consuming approach the ground truth for an object is somewhat adequately apparent. Machine learning is also leveraged by cosmologists to deduce quantities that include fundamental cosmological model's parameters or unobservable latent process.

An example of this is, for predicting the basic cosmological parameters based on the spatial distribution of dark matter, 3D convolutional neural networks are used (see Figure 4.4) (Ravanbakhsh et al., 2016; 2017). To complete this task, computationally intensive N-body simulations were used to train the networks, these simulations were for the developing of dark matter, specific values were assumed for the parameters (which are 10) found in the standard ΛCDM model. To apply this in a real-world setting, for visible matter, firstly there is a need to model the variance and bias of the visible tracers with regards to the underlying dark matter distribution. To reduce this gap, a convolutional neural network undergoes the training for learning a fast mapping between visible galaxies and dark matter (Zhang et al., 2019a; 2019b), this allowed for a compromise between computational cost and simulation accuracy. Being computationally expensive is a common challenge faced by this work as faced by other applications in lattice field theory, many-body quantum system, solid-state physics, and, due to this challenge there are adequately small numbers of statistically independent realizations of the simulations that are large X_μ (Zhang et al., 2018a; 2018b).

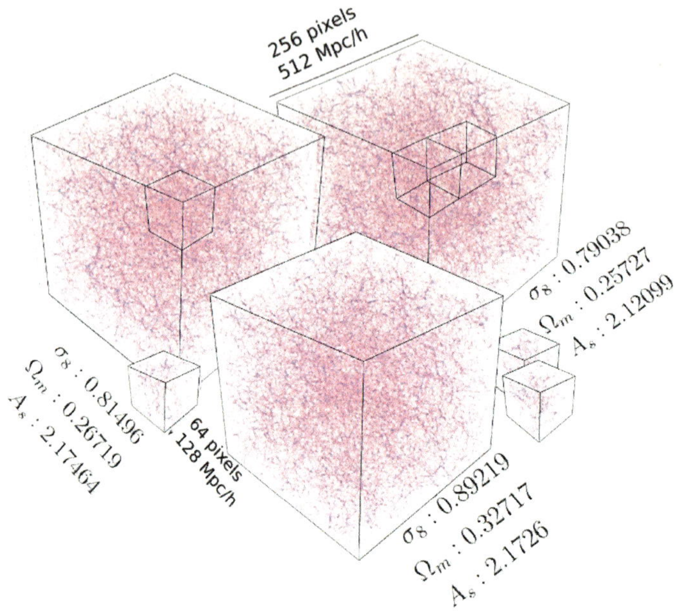

Figure 4.4. The image shows the distribution of dark matter in cubes. It is produced with the help of different sets of parameters. For prediction and training, each of the three cubes is divided into sub- cubes. In this figure, even though the cubes are produced by utilizing different cosmological parameters in the sampled (constrained) set, we cannot visually notice the effect (Ravanbakhsh et al., 2017).

4.5. PROBABILITY-FREE INFERENCE AND INVERSE PROBLEMS

As discussed earlier, both cosmology and particle physics are characterized by high-fidelity forward simulations. The simulations are either deterministic – as in N-body gravitational simulations or gravitational lensing case, or they are intrinsically stochastic – as in interactions found in simulations and probabilistic decay case. However, simulators of deterministic physics are even followed by observations about a probabilistic description, based on instrumental noise model or Poisson counts. In each of these cases, the simulation can be considered as the distribution p(X, Z| y) that is implicitly defined, where Z refers to the unobserved latent variables, taking random values, and X is the observed data, and the forward model's parameters are referred as y, such as ΛCDM cosmology's 10 parameters or the coefficients in a Lagrangian. We can characterize some scientific approaches as inverse problems, where y or Z can be inferred from X = x. We have considered the simplest classification cases, where regression and categorical values are taken by y where y ε ℝd. In scientific applications, uncertainty on the estimate is required whereas in general, the point estimates Z^(X = x) and y^(X = x) is useful.

The inverse problem's solution is ill-posed in many cases, as even the smallest of change in X results in huge changes in the estimate. Huge change implies a high variance in the estimator. In a few cases, a linear operator is equivalent to the forward model and we can express the probability (maximum one) estimate Z^MLE(X) or $y\hat{}_{MLE}(X)$ as matrix inversion. Under the case where the forward model is conditioned poorly, we relate the instability of the inverse to the matrix. The maximum probability estimate is likely to be high variance and it may be unbiased. Penalized maximum probability, Gaussian process regression, and ridge regression (Tikhonov regularization) are approaches that are related to the variance-bias trade-off.

In particle physics, this sort of problem is often termed as unfolding, in which one tends to look for the kinematic property's distribution of the collisions, and the quantity 's smeared version is represented by X. Similarly, inside the proton, we can estimate the proton density functions describing the gluons and quarks, and for this sort of problem, we can cast them as an inverse problem (Ball et al., 2015; Forte et al., 2002). In recent times, physically inspired kernels are applied to both Gaussian processes and neural networks for solving this problem (Bozson et al., 2018; Frate et al., 2017). In the cosmology's context, to denoise the time series of LIGO, Laser Interferometer Gravitational-Wave Observatory from a gravitational wave is an example of an inverse problem (Shen et al., 2017; 2019). GANs, Generative Adversarial Networks, are even used for inverse problems when de-noising and recovering of images of galaxies were required beyond naive deconvolution limits (Schawinski et al., 2017). There is another example that involves the estimation of a background object's image before be lensed gravitationally through a foreground object. Recently, Putzky and Welling, 2017 introduced recurrent inference machines for implicitly learning of such problems. To strong gravitational lensing, these machines are applied successfully (Morningstar et al., 2018, 2019).

An aspiring approach towards inverse problems includes providing probabilistic characterization in detail of y given X. One would desire for characterizing the function of probability L(y) = p(X = x| y) in frequentist paradigm, whereas the other one would desire for characterizing the posterior p(y| X = x) p(X = x| y)p(y) in a Bayesian formalism. The analogous situation takes place for latent variables inference, Z given X. Cosmology and particle physics both have approaches that are well-developed to statistical inference depending on the detailed modeling of the Hamiltonian Monte Carlo, Markov Chain Monte Carlo (MCMC), variational inference, and probability. However, traceability of the probability function is required by all of these approaches.

4.5.1. Probability-Free Inference

Somewhat surprisingly, probability or the probability density, p(X = x| y) that the simulator defines implicitly is often intractable. We can write the probability density, symbolically as p(X| y) = p(X, Z| y) dZ, here the Z represents the latent variables. The state-of-the-art of latent space of simulations is highly structured and massive, so we cannot analytically perform this integral. At the LHC, in single collision simulations, several hundred million components might be present in Z. The simulations are typically based on generated samples $(X_\mu, Z_\mu) \sim p(X, Z|y)$ and Monte Carlo techniques, from which we can estimate the density. It is hard to precisely estimate the densities when the X is high-dimensional. For instance, the estimation techniques of kernel density can be trusted around only 5-dimensions, and histogram-based naive approaches cannot be scaled to high dimensions, also, the distributions possess a huge dynamic range.

In addition to the field of cosmology and particle physics, the simulations implicitly define intractability of the probability is a primary issue in other fields of science as well, e.g., phylogenetics and epidemiology. It has led to develop an algorithm called probability-free inference: the ability for generating samples (in the forward mode) from the simulation is only required by it.

Approximate Bayesian Computation (ABC), is a prominent technique. In this technique, Bayesian inference is performed by using MCMC or an approach of rejection sampling, in which the probability approximates the probability p(ρ(X, x) < ε), where ρ(x,' x) is the distance metric between simulator's output x' and x itself, the observed data is referred as x, and tolerance parameter is referred as s. As $\varepsilon \rightarrow 0$, exact Bayesian inference is recovered: however, the procedure's efficiency vanishes. ABC faces a challenge for high-dimensional x, as it is the identification of the ρ(x' x) (distance measure), that reasonable acceptance efficiency is maintained by it without degrading the inference's quality(Beaumont et al., 2002; Marin et al., 2012; Marjoram et al., 2003; Sisson and Fan, 2011; Sisson et al., 2007). This estimating approach is similar to the conventional practice of using kernel density estimation or histograms for approximating p^(x| y) ~ p(x| y) in particle physics. In both cases, for reducing the data dimensionality, we require domain knowledge to identify useful summary. Universal probabilistic programming is utilized by an extension of the Approximate Bayesian Computation technique. Particularly, the inference compilation technique is an advanced form of importance sampling. In this technique, the random generation of numbers in the probabilistic program is controlled by a neural network that biases the simulation for producing x' (outputs) closer to the x (observed) (Le et al., 2017).

ABC is synonymously known as probability-free inference; however, other approaches involving learning probability ratio or an approximate probability (utilized as a surrogate), are present. For instance, neural density estimation with normalizing flows and autoregressive models are already used for this purpose (Larochelle and Murray, 2011; Papamakarios et al., 2017; Rezende and Mohamed, 2015) and for scaling to higher dimensional data (Cranmer and Louppe, 2016; Papamakarios et al., 2018). On other hand, training a classifier for discriminating between x ~ p(x|y') and x ~ p(x|y) can be utilized for estimating the probability ratio p(x|y)/p(x|y') ≈ r^(x|y, y'), that can further be utilized for inference in either Bayesian paradigm or the frequentist (Brehmer et al., 2018c; Cranmer et al., 2015; Hermans et al., 2019).

4.5.2. Examples in Particle Physics

Within particle physics, there are thousands of published results that include the Higgs boson discovery, involving the surrogate probability p^(x| y) based statistical inference that is constructed through the help of density estimation techniques: these techniques are applied to simulation generated synthetic datasets. The probability-free inference is relatively a new term, and for experimental particle physics, it is essential to its methodology. In recent times, the neural network-based suite of techniques of probability-free inference has not only been developed but also applied to physics models beyond the standard model presented in EFT abbreviated as effective field theory (Brehmer et al., 2018a,b). Around the standard model, a systematic expansion is provided by Effective field theory; coefficients parameterize the standard theory for quantum mechanical operators (the role of y is played by them in the setting). In this work, an interesting observation is that the joint score $t(x, z|\, y) = {}_{y} \log p(x, z|\, y)$ and the joint probability ratio $r(x, z|\, y, y')$ are tractable, though probability ratio and probability were difficult. We can use these tractable ratios for augmenting the training data, as Figure 4.5 shows, and it will result in dramatically improving the efficiency (sample) of these techniques (Brehmer et al., 2018c).

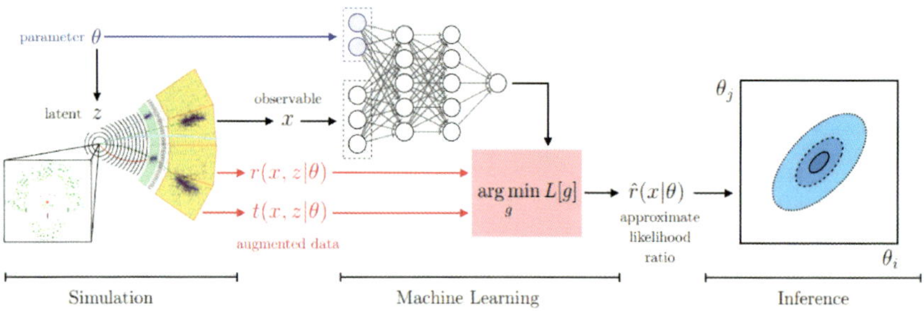

Figure 4.5. A representation of approaches based on machine learning to probability-free inference. For a neural network, training data is provided by the simulations, it is utilized for intractable probability and is utilized as a surrogate during inference (Brehmer et al., 2018b).

Moreover, a technique of inference compilation has been applied to a tau-lepton decay's inference. For this effort, a probabilistic programming protocol was developed that was capable of getting integrated into domain-specific simulation codes that already exist such as GEANT4 and SHERPA (Baydin et al., 2018; Casado et al., 2017). On the latent variables, Bayesian inference is provided by this approach, and as the posterior, deep interpretability corresponds to a distribution covering simulation's stack-traces, which allows probabilistic inspection of each aspect of the simulation.

For probability-free inference, AVO, adversarial variational optimization is another technique that has been developed (Louppe et al., 2019a). AVO parallels a generative network (adversarial) it is a generative model that is domain-specific simulation instead of a neural network. Rather than optimizing the network's parameters, this model aims the optimization of simulation's parameters, so that the target data distribution matches the generated data. Unlike neural networks, here the challenge is that mostly the simulators are not differentiable. For overcoming this issue, scientists have made the use of variational optimization techniques, as a differentiable surrogate loss function is provided by this technique. For tuning the simulation's parameters, this technique of optimization is being investigated. Bayesian optimization has also been utilized for this computationally intensive task (Ilten et al., 2017).

4.5.3. Examples in Cosmology

Within Cosmology, the scenario of constraining thick disk formation (Robin et al., 2014) of the Milky Way and at high redshift, morphological transformation rate of the galaxies (Cameron and Pettitt, 2012) are included in the early utilization of ABC, which aimed for tracking the Hubble parameter evolution. These experiences lead to developing CosmoABC like tools, which can help in streamlining the methodology application in cosmological applications (Ishida et al., 2015).

In recent times, motivated by the cosmology experiences, machine learning-based probability-free inference methods have been developed. While dealing with high-dimensional observations X, a compression strategy for data was developed for confronting the issues faced in the ABC method, this strategy learns summary statistics for maximizing the Fisher information (Alsing et al., 2018; Charnock et al., 2018). For the implicit probability, the sufficient statistics are approximated by the learned summary statistics in the neighborhood of some fiducial or nominal parameter value. Brehmer et al., 2018c is closely connected to this discussed approach. Recently, for learning summary statistics, robust to uncertainties (systematic), these approaches have been utilized (Alsing and Wandelt, 2019).

4.6. GENERATIVE MODELS

In machine learning, an active area of research involves the use of unsupervised learning for training a model that is capable of producing a distribution that is similar to some empirical distribution. This area of research includes normalizing flow models (Larochelle and Murray, 2011; Papamakarios et al., 2017; Rezende and Mohamed, 2015), autoregressive models, and generative adversarial networks (called as GANs) (Goodfellow et al., 2014), variational autoencoders (VAEs) (Kingma and Welling, 2013; Rezende et al., 2014). Interestingly, generative adversarial networks also face the same problem of the simulator defined intractability of the density, as faced by probability- free inference. GAN's density is intractable, hence a trick was needed, and on the other hand, if GAN's density were tractable, they would have been trained by using the method of standard maximum probability.

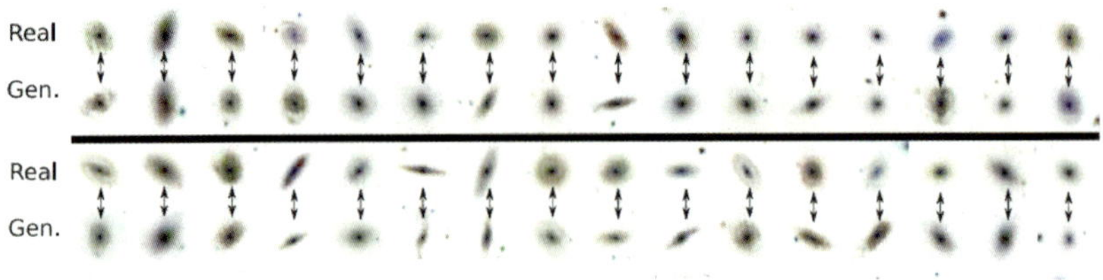

Figure 4.6. The GALAXY-ZOO dataset samples in comparison with the samples generated by the conditional generative adversarial network. The synthetic images are 128 colored images which are shown here in inverted form and are produced by conditioning different features y ε [0, 1]. In each column, the pair of generated and observed images correspond to the same value of y. The image is reproduced from (Ravanbakhsh et al., 2016).

Introducing an adversary is a trick, i.e., the discriminator network that is utilized for classifying the target distribution samples and the generative model samples. Between the two distributions, the probability ratio is effectively estimated by the discriminator, through which direct connections are provided to probability-free inference approaches that are based on classifiers (Cranmer and Louppe, 2016).

Operationally, a similar role is played by these models, as the one played by traditional scientific simulators. However, a causal model to underlie the data generation process is provided by traditional simulation codes. Nevertheless, traditional scientific simulators are extremely slow as the distributions emerge through a low-level microphysical description, e.g., at the LHC, simulating collisions involves physics (atomic-level) of scintillation and ionization. Similarly, in cosmology, the simulations involve gravitational interactions of different huge objects, and complex feedback processes, such as star formation and radiation, might also be included. Therefore, there is great importance to learn a fast approximation for the simulations.

In this direction, early work in particle physics includes GANs; it is used for energy deposits in calorimeters from particles (Paganini et al., 2018a,b); the ATLAS collaboration is studying this (ATLAS Collaboration, 2018). Within Cosmology, for cosmological structure formation, generative models are being used for learning the simulation (Rodríguez et al., 2018). For predicting the universe's structure formation (non-linear), a deep neural network was selected by an interesting hybrid approach. It predicted the formation from a residual, the physical simulation used was based on the theory of linear perturbation (He et al., 2018).

In other cases, there exist no well-motivated simulations or if they are present, they are impractical. Nonetheless, for the calibration purpose, it is valuable to have a generative model for this sort of data. Figure 4.6 shows an illustrative example coming from (Ravanbakhsh et al., 2016). It is pointed out that the cosmological survey's upcoming generation for weak gravitational lensing depends totally on accurate and precise measurements of the galaxies' apparent shapes. However, precise calibration is required by the shape measurement methods for fulfilling the requirements of accuracy that are needed for the science analysis. It is a challenging process as required high-quality galaxy images (a set of images) are expensive to collect. Hence, parametric bootstrap's implicit generalization is enabled by the GAN.

4.7. OUTLOOK AND CHALLENGES

Machine learning methods are being utilized by cosmology and particle physics for a long period, hence, the scope of topics involving the application of machine learning has increased significantly. It is now considered as a prime strategy for confronting the upgraded challenges of High-Luminosity LHC (Albertsson et al., 2018; Apollinari et al., 2015). Machine learning is also influencing the future strategies of experiments in both particle physics and cosmology (Ntampaka et al., 2019). A great amount of attention is gathered by one particular area that is the identification of the tracks that charged particles leave in an environment of high-luminosities (Farrell et al., 2018), it is being greatly focused.

All machine learning applied areas, for solving physics problems, desire for incorporating the basic domain knowledge in the shape of geometrical structures, compositional structure, hierarchical structure, or symmetries existing in the data- generation process or the data itself. Recently, in this direction, the machine learning community has done a spate of work (Bronstein et al., 2017; Cohen and Welling, 2016; Cohen et al., 2018, 2019; Kondor et al., 2018; Kondor and Trivedi, 2018). Physicists are closely following these developments and are incorporating them into contemporary research.

REFERENCES

1. Acciarri, R., Adams, C., An, R., Asaadi, J., Auger, M., Bagby, L., ... & Bishai, M. (2017). Convolutional neural networks applied to neutrino events in a liquid argon time projection chamber. Journal of Instrumentation, 12(03), P03011.
2. Albertsson, K., Altoe, P., Anderson, D., Anderson, J., Andrews, M., Espinosa, J. P. A., ... & Bonacorsi, D. (2018). Machine learning in high energy physics community white paper, 1, 1–33. arXiv preprint arXiv:1807.02876.
3. Alsing, J., & Wandelt, B. (2019). Nuisance hardened data compression for fast probability-free inference. Monthly Notices of the Royal Astronomical Society, 488(4), 5093–5103.
4. Alsing, J., Wandelt, B., & Feeney, S. (2018). Massive optimal data compression and density estimation for scalable, probability-free inference in cosmology. Monthly Notices of the Royal Astronomical Society, 477(3), 2874–2885.
5. Apollinari, G., Brüning, O., Nakamoto, T., & Rossi, L. (2017). High luminosity large hadron collider HL-LHC, 1–33. arXiv preprint arXiv:1705.08830.
6. Aurisano, A., Radovic, A., Rocco, D., Himmel, A., Messier, M. D., Niner, E., ... & Vahle, P. (2016). A convolutional neural network neutrino event classifier. Journal of Instrumentation, 11(09), P09001.
7. Baldi, P., Bauer, K., Eng, C., Sadowski, P., & Whiteson, D. (2016a). Jet substructure classification in high-energy physics with deep neural networks. Physical Review D, 93(9), 094034.
8. Baldi, P., Cranmer, K., Faucett, T., Sadowski, P., & Whiteson, D. (2016b). Parameterized machine learning for high-energy physics, 1, 1–33. arXiv preprint arXiv:1601.07913.
9. Baldi, P., Sadowski, P., & Whiteson, D. (2014). Searching for exotic particles in high-energy physics with deep learning. Nature communications, 5(1), 1–9.
10. Ball, R. D., Bertone, V., Carrazza, S., Deans, C. S., Del Debbio, L., Forte, S., ... & Ubiali, M. (2015). Parton distributions for the LHC Run II. Journal of High Energy Physics, 2015(4), 40.
11. Baydin, A. G., Heinrich, L., Bhimji, W., Gram-Hansen, B., Louppe, G., Shao, L., ... & Wood, F. (2018). Efficient Probabilistic Inference in the Quest for Physics Beyond the Standard Model.

12. Beaumont, M. A., Zhang, W., & Balding, D. J. (2002). Approximate Bayesian computation in population genetics. Genetics, 162(4), 2025–2035.
13. Benitez, N. (2000). Bayesian photometric redshift estimation. The Astrophysical Journal, 536(2), 571.
14. Bonnett, C., Troxel, M. A., Hartley, W., Amara, A., Leistedt, B., Becker, M. R., ... & Kind, M. C. (2015). Redshift distributions of galaxies in the DES Science Verification shear catalogue and implications for weak lensing, 1, 1–22. arXiv preprint arXiv:1507.05909.
15. Bozson, A., Cowan, G., & Spanò, F. (2018). Unfolding with Gaussian Processes, 1, 1–43. arXiv preprint arXiv:1811.01242.
16. Brammer, G. B., van Dokkum, P. G., & Coppi, P. (2008). EAZY: a fast, public photometric redshift code. The Astrophysical Journal, 686(2), 1503.
17. Brehmer, J., Cranmer, K., Louppe, G., & Pavez, J. (2018). A guide to constraining effective field theories with machine learning. Physical Review D, 98(5), 052004.
18. Brehmer, J., Cranmer, K., Louppe, G., & Pavez, J. (2018). Constraining effective field theories with machine learning. Physical Review Letters, 121(11), 111801.
19. Brehmer, J., Louppe, G., Pavez, J., & Cranmer, K. (2020). Mining gold from implicit models to improve probability-free inference. Proceedings of the National Academy of Sciences, 117(10), 5242–5249.
20. Breiman, L., Friedman, J. H., Olshen, R. A., & Stone, C. J. (1984). Classification and regression trees. Statistics/Probability Series, 1, 1–33.
21. Bronstein, M. M., Bruna, J., LeCun, Y., Szlam, A., & Vandergheynst, P. (2017). Geometric deep learning: going beyond euclidean data. IEEE Signal Processing Magazine, 34(4), 18–42.
22. Cameron, E., & Pettitt, A. N. (2012). Approximate Bayesian Computation for astronomical model analysis: a case study in galaxy demographics and morphological transformation at high redshift. Monthly Notices of the Royal Astronomical Society, 425(1), 44–65.
23. Carifio, J., Halverson, J., Krioukov, D., & Nelson, B. D. (2017). Machine learning in the string landscape. Journal of High Energy Physics, 2017(9), 157.

24. Carrasco Kind, M., & Brunner, R. J. (2013). TPZ: photometric redshift PDFs and ancillary information by using prediction trees and random forests. Monthly Notices of the Royal Astronomical Society, 432(2), 1483–1501.
25. Casado, M. L., Baydin, A. G., Rubio, D. M., Le, T. A., Wood, F., Heinrich, L., ... & Bhimji, W. (2017). Improvements to inference compilation for probabilistic programming in large-scale scientific simulators. arXiv preprint arXiv:1712.07901.
26. Charnock, T., Lavaux, G., & Wandelt, B. D. (2018). Automatic physical inference with information maximizing neural networks. Physical Review D, 97(8), 083004.
27. Choma, N., Monti, F., Gerhardt, L., Palczewski, T., Ronaghi, Z., Prabhat, P., ... & Bruna, J. (2018, December). Graph neural networks for icecube signal classification. In 2018 17th IEEE International Conference on Machine Learning and Applications (ICMLA) (Vol.1, pp. 386–391). IEEE.
28. Cohen, T. S., & Welling, M. (2016). Steerable cnns, 1, 1–33. arXiv preprint arXiv:1612.08498.
29. Cohen, T. S., Geiger, M., Köhler, J., & Welling, M. (2018). Spherical cnns, 1, 1–44. arXiv preprint arXiv:1801.10130.
30. Cohen, T. S., Weiler, M., Kicanaoglu, B., & Welling, M. (2019). Gauge equivariant convolutional networks and the icosahedral cnn, 1, 1–44. arXiv preprint arXiv:1902.04615.
31. Collaboration, A. T. L. A. S. (2018). Searches for heavy ZZ and ZW resonances in the ℓℓ qq and νν qq final states in pp collisions at√ s= 13 TeV with the ATLAS detector. Journal of High Energy Physics, 1803, 009.
32. Collett, T. E. (2015). The population of galaxy–galaxy strong lenses in forthcoming optical imaging surveys. The Astrophysical Journal, 811(1), 20.
33. Collister, A. A., & Lahav, O. (2004). ANNz: estimating photometric redshifts using artificial neural networks. Publications of the Astronomical Society of the Pacific, 116(818), 345.
34. Cranmer, K., & Louppe, G. (2016). Unifying generative models and exact probability-free inference with conditional bijections, 1, 1–33. J. Brief Ideas.

35. Cranmer, K., Pavez, J., & Louppe, G. (2015). Approximating probability ratios with calibrated discriminative classifiers, 1, 1–33. arXiv preprint arXiv:1506.02169.
36. de Oliveira, L., Kagan, M., Mackey, L., Nachman, B., & Schwartzman, A. (2016). Jet-images—deep learning edition. Journal of High Energy Physics, 2016(7), 69.
37. Duvenaud, D., Lloyd, J., Grosse, R., Tenenbaum, J., & Zoubin, G. (2013, February). Structure discovery in nonparametric regression through compositional kernel search. In International Conference on Machine Learning (Vol.1, pp. 1166–1174).
38. Estrada, J., Annis, J., Diehl, H. T., Hall, P. B., Las, T., Lin, H., ... & Tucker, D. (2007). A systematic search for high surface brightness giant arcs in a Sloan Digital Sky Survey cluster sample. The Astrophysical Journal, 660(2), 1176.
39. Feldmann, R., Carollo, C. M., Porciani, C., Lilly, S. J., Capak, P., Taniguchi, Y., ... & Aussel, H. (2006). The zurich extragalactic bayesian redshift analyzer and its first application: COSMOS. Monthly Notices of the Royal Astronomical Society, 372(2), 565–577.
40. Firth, A. E., Lahav, O., & Somerville, R. S. (2003). Estimating photometric redshifts with artificial neural networks. Monthly Notices of the Royal Astronomical Society, 339(4), 1195–1202.
41. Forte, S., s Garrido, L., Latorre, J. I., & Piccione, A. (2002). Neural network parametrization of deep-inelastic structure functions. Journal of High Energy Physics, 2002(05), 062.
42. Frate, M., Cranmer, K., Kalia, S., Vandenberg-Rodes, A., & Whiteson, D. (2017). Modeling smooth backgrounds and generic localized signals with gaussian processes, 1, 1–33. arXiv preprint arXiv:1709.05681.
43. Freund, Y., & Schapire, R. E. (1995, March). A desicion-theoretic generalization of on-line learning and an application to boosting. In European Conference on Computational Learning Theory (Vol.1, pp. 23–37). Springer, Berlin, Heidelberg.
44. Golkar, S., & Cranmer, K. (2018). Backdrop: Stochastic Backpropagation, 1, 1–33. arXiv preprint arXiv:1806.01337.
45. Goodfellow, I., Pouget-Abadie, J., Mirza, M., Xu, B., Warde-Farley, D., Ozair, S., ... & Bengio, Y. (2014). Generative adversarial nets. In Advances in Neural Information Processing Systems (Vol.1, pp. 2672–2680).

46. Guest, D., Collado, J., Baldi, P., Hsu, S. C., Urban, G., & Whiteson, D. (2016). Jet flavor classification in high-energy physics with deep neural networks. Physical Review D, 94(11), 112002.
47. Guest, D., Cranmer, K., & Whiteson, D. (2018). Deep learning and its application to LHC physics. Annual Review of Nuclear and Particle Science, 68, 161–181.
48. Hashimoto, K., Sugishita, S., Tanaka, A., & Tomiya, A. (2018). Deep learning and the AdS/CFT correspondence. Physical Review D, 98(4), 046019.
49. Henson, M. A., Barnes, D. J., Kay, S. T., McCarthy, I. G., & Schaye, J. (2016). The impact of baryons on massive galaxy clusters: halo structure and cluster mass estimates. Monthly Notices of the Royal Astronomical Society, stw2899, 1–44.
50. Hermans, J., Begy, V., & Louppe, G. (2019). Probability-free MCMC with approximate probability ratios, 1, 1–30. arXiv preprint arXiv:1903.04057, 10.
51. Hezaveh, Y. D., Levasseur, L. P., & Marshall, P. J. (2017). Fast automated analysis of strong gravitational lenses with convolutional neural networks. Nature, 548(7669), 555–557.
52. Ho, M., Rau, M. M., Ntampaka, M., Farahi, A., Trac, H., & Póczos, B. (2019). A Robust and Efficient Deep Learning Method for Dynamical Mass Measurements of Galaxy Clusters. The Astrophysical Journal, 887(1), 25.
53. Ilten, P., Williams, M., & Yang, Y. (2017). Event generator tuning using Bayesian optimization. Journal of Instrumentation, 12(04), P04028.
54. Ishida, E. E. O., Vitenti, S. D. P., Penna-Lima, M., Cisewski, J., de Souza, R. S., Trindade, A. M. M., ... & COIN collaboration. (2015). Cosmoabc: probability-free inference via population Monte Carlo approximate Bayesian computation. Astronomy and Computing, 13, 1–11.
55. Kasieczka, G., Plehn, T., Butter, A., Cranmer, K., Debnath, D., Dillon, B. M., ... & Gouskos, L. (2019). The machine learning landscape of top taggers, 1, 1–30. arXiv preprint arXiv:1902.09914.
56. Kingma, D. P., & Welling, M. (2013). Auto-encoding variational bayes, 1, 1–33. arXiv preprint arXiv:1312.6114.

57. Komiske, P. T., Metodiev, E. M., & Thaler, J. (2018). Energy flow polynomials: A complete linear basis for jet substructure. Journal of High Energy Physics, 2018(4), 13.
58. Komiske, P. T., Metodiev, E. M., & Thaler, J. (2019). Energy flow networks: deep sets for particle jets. Journal of High Energy Physics, 2019(1), 121.
59. Komiske, P. T., Metodiev, E. M., Nachman, B., & Schwartz, M. D. (2018). Learning to classify from impure samples with high-dimensional data. Physical Review D, 98(1), 011502.
60. Kondor, R., & Trivedi, S. (2018). On the generalization of equivariance and convolution in neural networks to the action of compact groups, 1, 1–33. arXiv preprint arXiv:1802.03690.
61. Kondor, R., Lin, Z., & Trivedi, S. (2018). Clebsch–gordan nets: a fully fourier space spherical convolutional neural network. In Advances in Neural Information Processing Systems (Vol.1, pp. 10117–10126).
62. Lanusse, F., Ma, Q., Li, N., Collett, T. E., Li, C. L., Ravanbakhsh, S., ... & Póczos, B. (2018). CMU DeepLens: deep learning for automatic image-based galaxy galaxy strong lens finding. Monthly Notices of the Royal Astronomical Society, 473(3), 3895–3906.
63. Larkoski, A. J., Moult, I., & Nachman, B. (2020). Jet substructure at the Large Hadron Collider: a review of recent advances in theory and machine learning. Physics Reports, 841, 1–63.
64. Larochelle, H., & Murray, I. (2011, June). The neural autoregressive distribution estimator. In Proceedings of the Fourteenth International Conference on Artificial Intelligence and Statistics (Vol.1, pp. 29–37).
65. Le, T. A., Baydin, A. G., & Wood, F. (2017, April). Inference compilation and universal probabilistic programming. In Artificial Intelligence and Statistics (Vol.1, pp. 1338–1348).
66. Leistedt, B., Hogg, D. W., Wechsler, R. H., & DeRose, J. (2019). Hierarchical modeling and statistical calibration for photometric redshifts. The Astrophysical Journal, 881(1), 80.
67. Li, N., Gladders, M. D., Rangel, E. M., Florian, M. K., Bleem, L. E., Heitmann, K., ... & Fasel, P. (2016). Pics: Simulations of strong gravitational lensing in galaxy clusters. The Astrophysical Journal, 828(1), 54.
68. Likhomanenko, T., Ilten, P., Khairullin, E., Rogozhnikov, A., Ustyuzhanin, A., & Williams, M. (2015, December). LHCb topological

trigger reoptimization. In J. Phys. Conf. Ser (Vol. 664, No. 8, p. 082025).

69. Louppe, G., Cho, K., Becot, C., & Cranmer, K. (2019a). QCD-aware recursive neural networks for jet physics. Journal of High Energy Physics, 2019(1), 57.
70. Louppe, G., Hermans, J., & Cranmer, K. (2019b, April). Adversarial variational optimization of non-differentiable simulators. In The 22nd International Conference on Artificial Intelligence and Statistics (pp. 1438–1447).
71. Louppe, G., Kagan, M., & Cranmer, K. (2016). Learning to pivot with adversarial networks, 1, 1–33. arXiv preprint arXiv:1611.01046.
72. Marin, J. M., Pudlo, P., Robert, C. P., & Ryder, R. J. (2012). Approximate Bayesian computational methods. Statistics and Computing, 22(6), 1167–1180.
73. Marjoram, P., Molitor, J., Plagnol, V., & Tavaré, S. (2003). Markov chain Monte Carlo without likelihoods. Proceedings of the National Academy of Sciences, 100(26), 15324–15328.
74. Marshall, P. J., Hogg, D. W., Moustakas, L. A., Fassnacht, C. D., Bradač, M., Schrabback, T., & Blandford, R. D. (2009). Automated detection of galaxy-scale gravitational lenses in high-resolution imaging data. The Astrophysical Journal, 694(2), 924.
75. Metodiev, E. M., Nachman, B., & Thaler, J. (2017). Classification without labels: Learning from mixed samples in high energy physics. Journal of High Energy Physics, 2017(10), 174.
76. Morningstar, W. R., Hezaveh, Y. D., Levasseur, L. P., Blandford, R. D., Marshall, P. J., Putzky, P., & Wechsler, R. H. (2018). Analyzing interferometric observations of strong gravitational lenses with recurrent and convolutional neural networks, 1, 1–33. arXiv preprint arXiv:1808.00011.
77. Morningstar, W. R., Levasseur, L. P., Hezaveh, Y. D., Blandford, R., Marshall, P., Putzky, P., ... & Welling, M. (2019). Data-driven reconstruction of gravitationally lensed galaxies using recurrent inference machines. The Astrophysical Journal, 883(1), 14.
78. Ntampaka, M., Trac, H., Sutherland, D. J., Battaglia, N., Póczos, B., & Schneider, J. (2015). A machine learning approach for dynamical mass measurements of galaxy clusters. The Astrophysical Journal, 803(2), 50.

79. Ntampaka, M., Trac, H., Sutherland, D. J., Fromenteau, S., Póczos, B., & Schneider, J. (2016). Dynamical mass measurements of contaminated galaxy clusters using machine learning. The Astrophysical Journal, 831(2), 135.
80. Ntampaka, M., ZuHone, J., Eisenstein, D., Nagai, D., Vikhlinin, A., Hernquist, L., ... & Torrey, P. (2019). A Deep Learning Approach to Galaxy Cluster X-Ray Masses. The Astrophysical Journal, 876(1), 82.
81. Paganini, M., de Oliveira, L., & Nachman, B. (2018a). Accelerating science with generative adversarial networks: an application to 3D particle showers in multilayer calorimeters. Physical review letters, 120(4), 042003.
82. Paganini, M., de Oliveira, L., & Nachman, B. (2018b). CaloGAN: Simulating 3D high energy particle showers in multilayer electromagnetic calorimeters with generative adversarial networks. Physical Review D, 97(1), 014021.
83. Papamakarios, G., Pavlakou, T., & Murray, I. (2017). Masked autoregressive flow for density estimation. In Advances in Neural Information Processing Systems (Vol.1, pp. 2338–2347).
84. Papamakarios, G., Sterratt, D., & Murray, I. (2019, April). Sequential neural probability: Fast probability-free inference with autoregressive flows. In The 22nd International Conference on Artificial Intelligence and Statistics (Vol.1, pp. 837–848).
85. Peel, A., Lalande, F., Starck, J. L., Pettorino, V., Merten, J., Giocoli, C., ... & Baldi, M. (2019). Distinguishing standard and modified gravity cosmologies with machine learning. Physical Review D, 100(2), 023508.
86. Poczos, B., Schneider, J., Brandenburg, A., Kahniashvili, T., Tevzadze, A., & Trac, H. CDS&E: Collaborative Research: Machine Learning for Automated Discovery and Control in Turbulent Plasma, 1, 1–33.
87. Putzky, P., & Welling, M. (2017). Recurrent inference machines for solving inverse problems, 1, 1–22. arXiv preprint arXiv:1706.04008.
88. Radovic, A., Williams, M., Rousseau, D., Kagan, M., Bonacorsi, D., Himmel, A., ... & Wongjirad, T. (2018). Machine learning at the energy and intensity frontiers of particle physics. Nature, 560(7716), 41–48.
89. Ravanbakhsh, S., Lanusse, F., Mandelbaum, R., Schneider, J., & Poczos, B. (2017, February). Enabling dark energy science with deep

generative models of galaxy images. In Thirty-First AAAI Conference on Artificial Intelligence, 31, 1–44.

90. Ravanbakhsh, S., Oliva, J., Fromenteau, S., Price, L., Ho, S., Schneider, J., & Póczos, B. (2016, June). Estimating cosmological parameters from the dark matter distribution. In International Conference on Machine Learning (Vol.1, pp. 2407–2416).
91. Ren, S., He, K., Girshick, R., & Sun, J. (2015). Faster r-cnn: Towards real-time object detection with region proposal networks. In Advances in Neural Information Processing Systems (Vol.1, pp. 91–99).
92. Rezende, D. J., & Mohamed, S. (2015). Variational inference with normalizing flows, 1, 1–33. arXiv preprint arXiv:1505.05770.
93. Rezende, D. J., Mohamed, S., & Wierstra, D. (2014). Stochastic backpropagation and approximate inference in deep generative models, 1, 1–33. arXiv preprint arXiv:1401.4082.
94. Robin, A. C., Reylé, C., Fliri, J., Czekaj, M., Robert, C. P., & Martins, A. M. M. (2014). Constraining the thick disc formation scenario of the Milky Way. Astronomy & Astrophysics, 569, A13.
95. Rodríguez, A. C., Kacprzak, T., Lucchi, A., Amara, A., Sgier, R., Fluri, J., ... & Réfrégier, A. (2018). Fast cosmic web simulations with generative adversarial networks. Computational Astrophysics and Cosmology, 5(1), 4.
96. Roe, B. P., Yang, H. J., Zhu, J., Liu, Y., Stancu, I., & McGregor, G. (2005). Boosted decision trees as an alternative to artificial neural networks for particle identification. Nuclear Instruments and Methods in Physics Research Section A: Accelerators, Spectrometers, Detectors and Associated Equipment, 543(2–3), 577–584.
97. Rogozhnikov, A., Bukva, A., Gligorov, V., Ustyuzhanin, A., & Williams, M. (2015). New approaches for boosting to uniformity. Journal of Instrumentation, 10(3), T03002.
98. Schawinski, K., Zhang, C., Zhang, H., Fowler, L., & Santhanam, G. K. (2017). Generative adversarial networks recover features in astrophysical images of galaxies beyond the deconvolution limit. Monthly Notices of the Royal Astronomical Society: Letters, 467(1), L110-L114.
99. Shanahan, P. E., Trewartha, D., & Detmold, W. (2018). Machine learning action parameters in lattice quantum chromodynamics. Physical Review D, 97(9), 094506.

100. Shen, H., George, D., Huerta, E. A., & Zhao, Z. (2019, May). Denoising gravitational waves with enhanced deep recurrent denoising auto-encoders. In ICASSP 2019–2019 IEEE International Conference on Acoustics, Speech, and Signal Processing (ICASSP) (Vol.1, pp. 3237–3241). IEEE.
101. Shimmin, C., Sadowski, P., Baldi, P., Weik, E., Whiteson, D., Goul, E., & Søgaard, A. (2017). Decorrelated jet substructure tagging using adversarial neural networks. Physical Review D, 96(7), 074034.
102. Sisson, S. A., & Fan, Y. (2011). Probability-free MCMC (Vol. 1, pp. 313–335). Chapman & Hall/CRC, New York.
103. Sisson, S. A., Fan, Y., & Tanaka, M. M. (2007). Sequential Monte Carlo without likelihoods. Proceedings of the National Academy of Sciences, 104(6), 1760–1765.
104. Stevens, J., & Williams, M. (2013). uBoost: A boosting method for producing uniform selection efficiencies from multivariate classifiers. Journal of Instrumentation, 8(12), P12013.
105. Z Zhang, P., Shen, H., & Zhai, H. (2018b). Machine learning topological invariants with neural networks. Physical Review Letters, 120(6), 066401.
106. Zhang, L., Han, J., Wang, H., Car, R., & Weinan, E. (2018a). Deep potential molecular dynamics: a scalable model with the accuracy of quantum mechanics. Physical Review Letters, 120(14), 143001.
107. Zhang, L., Lin, D. Y., Wang, H., Car, R., & Weinan, E. (2019a Active learning of uniformly accurate interatomic potentials for materials simulation. Physical Review Materials, 3(2), 023804.
108. Zhang, W., Wang, L., & Wang, Z. (2019b). Interpretable machine learning study of the many-body localization transition in disordered quantum Ising spin chains. Physical Review B, 99(5), 054208.

Chapter 5

Machine Learning in Artificial Intelligence

CONTENTS

5.1. INTRODUCTION

In April 2018, Mark Zuckerberg emphasized the essential proficiency of Facebook's "terrorist propaganda" or "AI tools to identify hate speech" in a hearing of the US senate (Gupta, 2017). Such tasks would be usually described by researchers as *classification tasks* within the area of *(supervised) machine learning* in which particular instances are recognized within social media platforms (Becker et al., 200; McCarthy et al., 2006; Chavan, 2020). Nonetheless, the terminology *artificial intelligence (AI)* is generally used in place of machine learning because of the growth of AI recognition – not only by the CEO of Facebook as given in the above example or other talks but across numerous application-oriented and theoretical contributions in the latest literature as well (Singer et al., 2000; Chen & Van Beek, 2001; Hong, 2001). There is still a statement from Peng and Zhang (2007) that he still put into practice the AI term interchangeably with machine learning even though aware it is not correct. Although such uncertainty may results in multiple ambiguousness when talking about concepts, methods, and results both in practice and research (Zhou et al., 2007; Cioffi et al., 2020).

It seems interesting that there is barely any beneficial scientific description regardless of the repeated use of these terms. Hence, the target of this chapter is to highlight the connection of two terms *artificial intelligence* and *machine learning*. The role of machine learning is elaborated within the instantiations of artificial intelligence, specifically within intelligent agents. Therefore, the perspective of machine learning is taken on the competence of intelligent agents along with the corresponding implementation.

5.2. RELATED WORK

Various concepts, notions, or definitions of artificial intelligence and machine learning are reviewed as a base for our conceptual work within research in existence. Additionally, the theories are elaborated in greater detail which plays an important part in our framework.

5.2.1. Terminology

Artificial intelligence and machine learning, plus the terms deep learning, statistical learning, and data mining are associated, usually cited in the same context and occasionally used interchangeably. As the terms are generally known in differing communities, their exact usage and significance vary broadly.

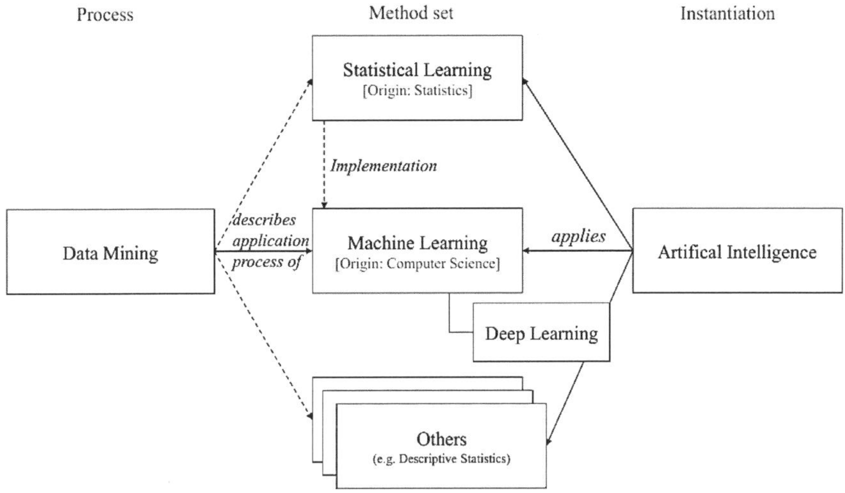

Figure 5.1. General terminology.

[Source:https://www.researchgate.net/publication/327802544_Machine_Learning_in_Artificial_Intelligence_Towards_a_Common_Understanding.]

For example, there is a focus on *statistical learning* in the area of statistics, which is described as a set of algorithms and methods to predict results, gain knowledge, and make decisions by setting up models from a data set (Halal, 2003; Wang et al., 2007). Machine learning can be viewed as an implementation of statistical learning from a statistics perspective (Vukasin, 1998).

The focus of *machine learning* within the line of computer science is to design efficient algorithms to deal with problems with computational resources (Metaxiotis et al., 2004). Even though approaches from statistics are utilized by machine learning, but some methods are also included that are not thoroughly based on past work of statisticians—leading to well-cited and contemporary contributions to the field (Raynor & Shoup, 1999; Stefanuk & Zhozhikashvili, 2002). Since the previous years, the method of deep learning specifically produced great interest (Tay & Ho, 1992). *Deep learning* models are built upon multiple processing layers that are adequate for learning descriptions of data with different levels of abstraction. The proficiency of machine learning has been remarkably improved by deep learning, such as in image or speech recognition (Wongpinunwatana et al., 2000; Oke, 2008).

In delimitation to the prior terms, the mechanism to apply quantitative analytical methods is described by *data mining*, which aids to solve real-world problems, like in business settings (Carvalho et al., 2019). With a machine learning perspective, data mining is the procedure of producing purposeful machine learning models. The objective is to apply machine learning algorithms to data to acquire intuition instead of establishing additional know-how about them. Therefore, machine learning is considered as a base for data mining (Markham et al., 2000).

Whereas, the approaches such as statistical learning, machine learning, or alternate approaches like descriptive statistics are applied by *artificial intelligence* to imitate intelligence in machines.

The base of the rest of this work is laid by the terms outlined in this paragraph and Figure 5.1. Nonetheless, there is a controversial discussion on relationships of the concepts and overall terminology (Cortes & Vapnik, 1995; Kotsiantis et al., 2007). Hence, this paper has focused on understanding terminology and more specifically, to clear up the part of machine learning within AI. To attain an extensive understanding of the terminologies AI and machine learning, both are additionally examined in detail.

5.2.2. Machine Learning

Even though while learning process, there is a use of statistical methods, programming of rules or approaches to solving a problem or a manual adjustment is not demanded. More specifically, the aim of (supervised) machine learning techniques is always to construct a model by employing an algorithm on a set of known data elements to achieve knowledge of an unknown set of data (Kitchenham, 2004; Duan et al., 2019).

So, the methods of "creation" of a machine learning model marginally differ to their definition of stages but generally exploit three main stages (1) model initiation, (2) performance estimation, and (3) deployment (Larranaga et al., 2005; De Felice et al., 2018):

In the model initiation stage, a problem is defined by human users; a dataset is prepared and processed, and an appropriate machine learning algorithm is selected for the given job. In the performance estimation stage, validation of numerous parameter permutations expressing the algorithm and a selection of well-performing configuration corresponding to its performance to solve a particular task is performed. In the end, deployment of the model is accomplished and makes it functional to perform on unknown or unseen data.

In general, learning represents a key aspect of a human's cognition which directs to all processes by which the sensory input is reduced, transformed, stored, elaborated, used, and recovered (Krawczyk, 2016; Wuest et al., 2016). A wide range of information is processed by humans by taking advantage of abstract knowledge that assists to better figure out upcoming input. Machine learning models are capable of imitating the cognitive strength of a human being in an isolated fashion because of their adaptive nature.

Still, a set of methods are merely represented by machine learning that allows learning patterns in current data, hence producing analytical models that can be used inside larger IT commodity.

5.2.3. Artificial Intelligence

The subject of artificial intelligence (AI) is implanted in various research disciplines, e.g., philosophy, computer science, or futures studies. This work is chiefly focused on the area of computer science because it is the most appropriate one in recognizing the participation of machine learning to AI and in differentiating both terminologies (Pérez-Ortiz et al., 2016; Dutton, 2018).

AI research can be divided into divergent research streams. These streams diverge on the one hand as to the intention of AI application (*acting* vs. *thinking*), on the other hand as to the sort of decision making (aiming a *rational decision, human-like decision* vs. an *ideal*). This discrepancy results in four research currents that are illustrated in Figure 5.2 (Lieber et al., 2013; Sachs et al., 2019).

Under the "Cognitive Modeling" (that is humanly thinking) stream, an AI should be a machine having a mind (Ghahramani, 2015). This involves achieving human thinking as well, such that when giving the same input to the machine as given to the human, it would return the same output as human returns. Not only that, but it would also give the same reasoning steps that would lead to the conclusion (Dunjko & Briegel, 2017; 2018).

The "Laws of Thought" stream (that is thinking rationally) needs an AI to reach the rational decision regardless of what a human might respond.

Thus, the laws of thought must be followed by an AI with the help of computational models that manifest logic (Kibria et al., 2018; Syam & Sharma, 2018).

Objective / Application to	Humanly	Rationally
Thinking	Cognitive Modeling	"Laws of thought "
Acting	Turing Test	Rational Agent

Figure 5.2. Research streams of AI-based on Russell and Norvig.

[Source:https://www.researchgate.net/publication/327802544_Machine_Learning_in_Artificial_Intelligence_Towards_a_Common_Understanding.]

The "Turing Test" stream indicates if an AI takes a move "smartly," "humanly," while communicating with humans. To achieve these functions, an AI must accomplish human tasks at least as fine as humans. Turing Test can test these requirements (Dimiduk et al., 2018; Gandhi et al., 2018).

Lastly, the "Rational Agent" stream deems an AI being a rational or smart agent. This agent acts independently intending to produce rationally ideal results.

Another way to describe AI is by representing intelligence in general and to manufacture intelligent machines through resulting insights. Theories of human intelligence, psychological definitions, and intelligence tests are used by Arel et al. (2010) to express a metric of intelligence. Based on their description, an agent-environment framework is used by them to represent intelligence in general and if the agent is a machine then to represent artificial intelligence especially. There are many similarities in their framework and "acting rationally" stream. Despite expressing AI in general, there is another topic in the area of AI research, i.e., classification of AI. An *artificial general intelligence* (AGI) is an AI that takes a step at least on a similar level as a human brain in any domain, but without needing consciousness. While, a *narrow AI* is an AI that surpasses or rivals the human brain just, in particular, restricted tasks. In succeeding, the "Rational Agent" stream is discussed in more detail because of its importance when concerning the implementation of machine learning within artificial intelligence. The other three research streams are discussed in section 3 and shown the compatibility of them with our framework of an agent-based AI.

The acting of agents manifests the intelligence itself under the "Rational Agent" stream. There are five characteristics of these agents, i.e., they perceive their environment, operate autonomously, adapt to change, persist over a prolonged period, and create and pursue goals. The actions are defined by an agent for itself as well as with an environment with which it interacts. Sensors are used to recognize the environment, the agent program decides the actions to be performed with input data, and its actuators perform that action. According to this performance measure, the agent should act based on the feasible actions and the recent and prior knowledge of the environment to accomplish the highest anticipated outcome to be a rational agent. According to Russel & Norvig, there are four diverse types of agents that the agent program segments for a general division of agents that are: (i) a *simple reflex* agent that reacts respective on its sensor data only, (ii) a *model-based reflex* agent which is recognized as an internal state of the agent as well, (iii) a *goal-based* agent that takes best decision to reach its goals. It gives a binary decision that indicates either a goal is fulfilled or not; and (iv) a *utility-based* agent that has a utility function instead of a binary goal that it tries to escalate. By extending the agent's program, it can be considered as a *learning agent*. A performance element is incorporated by such a *learning agent* that decides an action based on the sensor data and a learning element, which generates own problems, gets responses from the environment, and improves the performance element possibly. There are three components of the agent-environment framework (i) an agent, (ii) an environment, and (iii) a goal. Intelligence is the metric to measure the agent's competency to accomplish goals in a broad range of environments. The input is given to the agent by perceptions created from the environment. One kind of perception is reward signals that represent how well the goals of the agent are accomplished, whereas others are the observations of the environment. The decision is taken by the agent based on these input signals to perform actions that are returned to the environment as signals.

5.3. A FRAMEWORK FOR UNDERSTANDING THE ROLE OF MACHINE LEARNING IN ARTIFICIAL INTELLIGENCE

Our concept is based on the framework of Russel & Norvig (2003) to interpret the interaction of AI and machine learning. Their foundation is laid on the differentiation between the two aims of AI application, i.e., *thinking* and *acting*.

5.3.1. Layers of agenTs

A perspective is needed to be taken that has a spotlight on the implementation of intelligent agents to understand the part of machine learning within AI. This perspective is required because it allows to map a variety of components and tasks of machine learning to the abilities of intelligent agents. Consider the abilities of *acting* and *thinking* of an intelligent agent and transforming this into the provision of software design, then *acting* abilities are considered as a *frontend*, whereas the *thinking* part is considered as *backend.* Function and form are generally separated by software engineers strictly that permit for more independence and flexibility and also enable parallel development. The interface by which the environment interacts is known as frontend that can endure many forms. For intelligent agents, it can be a human-readable application, a very abstract, machine-readable web interface, or even a humanoid template with expanded expression proficiency. Two technical components are required for the frontend to reach out to the environment, i.e., actuators and sensors. The changes or events of the environment are detected and forwarded to the backend via frontend by *sensors.* For example, they can read visuals of an interplay with a human, and the temperature inside an industrial production machine. On the other hand, *actuators* are responsible for controlling and moving a mechanism. In short, sensors process information, while actuators *act*, e.g., by changing the facial expressions of a humanoid or automatically buying stocks. It can say that with the frontend, the Turing test happens at the interplay of the environment, more specifically is the combo of sensors and actuators for the examination of the agent's AI of *acting humanly*. Regardless of each frontend including sensors and actuators, our focus is not on the appearance of frontend; the significance is the existence of an encapsulated frontend, a backend-independent.

The essential functionalities are provided by the backend that portrays the *thinking* capacities of an intelligent agent. Hence, the agent should learn and practice learned knowledge.

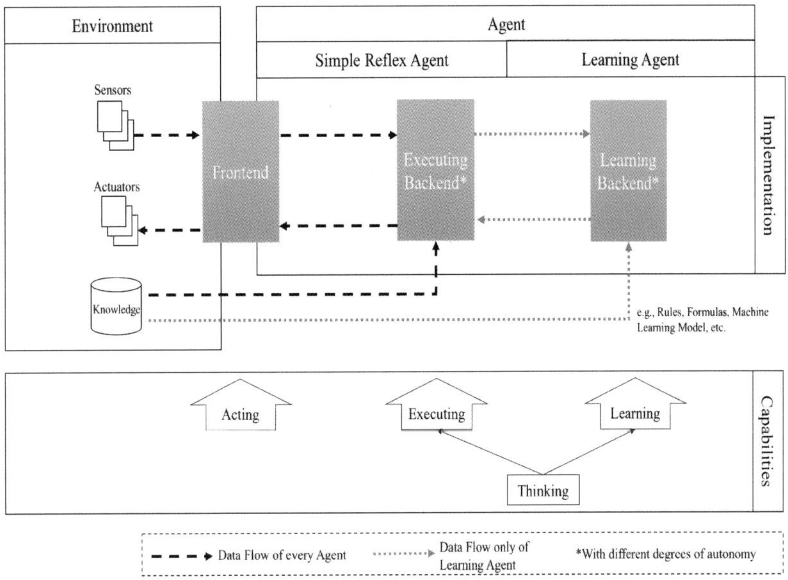

Figure 5.3. Conceptual framework of artificial intelligence and machine learning.

[Source:https://www.researchgate.net/publication/327802544_Machine_Learning_in_Artificial_Intelligence_Towards_a_Common_Understanding.]

As a result, this implementation layer is relevant to machine learning. With the perspective of supervised machine learning, the process of constructing (=training) adequate machine learning models, and the process of executing deployed models should be differentiated. Thus, the *thinking* layer is refined into a *learning* sublayer (model building) and also an *executing* sublayer (model execution) to understand further the part of machine learning in intelligent agents. Therefore, the essential implementation for learning sublayer is considered as a *learning backend*, while executing a sublayer is indicated by *executing backend*.

5.3.2. Types of learning

The learning backend prescribes *if* the intelligent agent is ready to learn, and then *how* it can learn, e.g., what kind of data processing is used, which exact algorithms it utilizes, how concept drift is managed, etc. So, two different kinds of intelligent agents are selected based on the terms given by Russel & Norvig (2003) (i) *simple-reflex agents*, and (ii) *learning agents*. This distinction holds especially for the machine learning viewpoint on AI.

Because it views either the latent models in the *thinking* layer are trained once and never have to do with again (i.e., simple-reflex)—or repeatedly adaptive and updated (i.e., learning). Appropriate examples for both can be taken from recent literature.

As an instance for simple-reflex agents, an early alarming system of pneumonia was build and deployed by Oroszi and Ruhland in hospitals: After deployment, the adaptive learning of the system might be crucial during constructing and evaluating the model for the agent gives promising results. In various areas, other agents having single-trained models are generally known, e.g., prediction of pedestrians, object annotation, or for anaphora resolutions. In contrast, examples of learning agents are also found in recent literature. The concept "never-ending learning" agents have been presented by Mitchell et al. that have an intense focus on continuously constructing and amending models within agents. In general, the selection of these characteristics (i.e., simple-reflex vs. learning agent) impacts the entire design of the agent and also the offering of machine learning. Figure 5.3 illustrates the sketch of our resulting framework. Concluding that machine learning contributes as a once-trained model within the execution sublayer for the simple-reflex agent. While it participates in the learning sublayer of a learning agent to frequently enhance the model within the execution sublayer. This enhancement is dependent on feedback and knowledge, which is deducted from the environment through the execution layer.

5.3.3. The continuum Between Human Involvement and Machine Involvement

There is a sign that how much automated the vital processes are for executing and learning backend instead of if and how elemental machine learning models are bought up to date. Numerous process steps are involved in each machine learning task including: (1) selection of data source, (2) data gathering, (3) preprocessing, (4) model construction, (5) evaluation, (6) deployment, (7) execution, and (8) improvement. Discussion of individual steps is not in the scope of this paper, the particular interest in each essential task of the machine learning lifecycle is the automation and autonomy of these tasks as an implementation in the agent.

For example, as long as the execution of a once-built model can be quite easily automated, the automated identification or retraining of a sufficient data source for a new problem, and also a construction of the self-induced model is more difficult. Consequently, human involvement is necessary to consider in the machine learning activities of an intelligent agent, as illustrated in Figure 5.4.

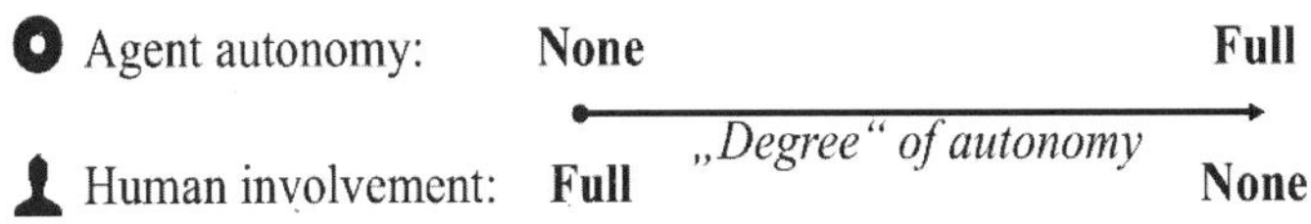

Figure 5.4. Degree of human involvement and agent autonomy.

[Source: https://www.researchgate.net/publication/327802544_Machine_Learning_in_Artificial_Intelligence_Towards_a_Common_Understanding]

Even though it is difficult to mark a sharp line between all probable formats of human intervention in the machine learning tasks of an intelligent agent, this phenomenon is seen as a continuum. The range of continuum on one extreme is none or slight agent autonomy with full human intervention, and on the other extreme full agent autonomy and none or slight human intervention for the delivered task. For instance, the task of driving a car autonomously by an intelligent agent while taking the traffic signs into account already shows a high degree of agent autonomy. Nevertheless, if a new traffic sign confronts the agent, human intervention might still requires the learning of this new occurrence because the agent might not be able to learn thoroughly by itself. While characterizing AI and the underlying machine learning models, the major interest is in the paramount interference of humans, particularly in the *thinking* layer (= executing backend & learning backend). For every step of machine learning, the degree of autonomy can be examined and may support to distinguish the autonomy of an agent regarding the related machine learning tasks.

REFERENCES

1. Abajian, A., Murali, N., Savic, L. J., Laage-Gaupp, F. M., Nezami, N., Duncan, J. S., ... & Chapiro, J. (2018). Predicting treatment response to intra-arterial therapies for hepatocellular carcinoma with the use of supervised machine learning—an artificial intelligence concept. *Journal of Vascular and Interventional Radiology*, *29*(6), 850–857.
2. Arel, I., Rose, D. C., & Karnowski, T. P. (2010). Deep machine learning-a new frontier in artificial intelligence research [research frontier]. *IEEE computational intelligence magazine*, *5*(4), 13–18.
3. Becker, A., Bar-Yehuda, R., & Geiger, D. (2000). Randomized algorithms for the loop cutset problem. *Journal of Artificial Intelligence Research*, *12*, 219–234.
4. Carvalho, T. P., Soares, F. A., Vita, R., Francisco, R. D. P., Basto, J. P., & Alcalá, S. G. (2019). A systematic literature review of machine learning methods applied to predictive maintenance. *Computers & Industrial Engineering*, *137*, 106024.
5. Chavan, C. U. (2020). Artificial Intelligence and Smart Machine Learning: Progress, Trends and Directions. *Studies in Indian Place Names*, *40*(38), 626–635.
6. Chen, X., & Van Beek, P. (2001). Conflict-directed backjumping revisited. *Journal of Artificial Intelligence Research*, *14*, 53–81.
7. Cioffi, R., Travaglioni, M., Piscitelli, G., Petrillo, A., & De Felice, F. (2020). Artificial intelligence and machine learning applications in smart production: Progress, trends, and directions. *Sustainability*, *12*(2), 492.
8. Cortes, C., & Vapnik, V. (1995). Support-vector networks. *Machine Learning*, *20*(3), 273–297.
9. De Felice, F., Petrillo, A., & Zomparelli, F. (2018). Prospective design of smart manufacturing: An Italian pilot case study. *Manufacturing Letters*, *15*, 81–85.
10. Dimiduk, D. M., Holm, E. A., & Niezgoda, S. R. (2018). Perspectives on the impact of machine learning, deep learning, and artificial intelligence on materials, processes, and structures engineering. *Integrating Materials and Manufacturing Innovation*, *7*(3), 157–172.

11. Duan, Y., Edwards, J. S., & Dwivedi, Y. K. (2019). Artificial intelligence for decision making in the era of Big Data–evolution, challenges and research agenda. *International Journal of Information Management*, *48*, 63–71.
12. Dunjko, V., & Briegel, H. J. (2017). Machine learning\& artificial intelligence in the quantum domain, 1, 1–33. *arXiv preprint arXiv:1709.02779*.
13. Dunjko, V., & Briegel, H. J. (2018). Machine learning & artificial intelligence in the quantum domain: a review of recent progress. *Reports on Progress in Physics*, *81*(7), 074001.
14. Dutton, T. (2018). An overview of national AI strategies. *Medium. Politics+ AI. June, 28, 1–33.*
15. Gandhi, S., Mosleh, W., Shen, J., & Chow, C. M. (2018). Automation, machine learning, and artificial intelligence in echocardiography: a brave new world. *Echocardiography*, *35*(9), 1402–1418.
16. Ghahramani, Z. (2015). Probabilistic machine learning and artificial intelligence. *Nature*, *521*(7553), 452–459.
17. Gupta, N. (2017). A Literature Survey on Artificial Intelligence. *International Journal Of Engineering Research & Technology (IJERT) ICPCN*, *5*(19), 1–33.
18. Halal, W. E. (2003) Artificial intelligence is almost here. *On the Horizon, 1, 1–44.*
19. Hong, J. (2001). Goal recognition through goal graph analysis. *Journal of Artificial Intelligence Research*, *15*, 1–30.
20. Kibria, M. G., Nguyen, K., Villardi, G. P., Zhao, O., Ishizu, K., & Kojima, F. (2018). Big data analytics, machine learning, and artificial intelligence in next-generation wireless networks. *IEEE access*, *6*, 32328–32338.
21. Kitchenham, B. (2004). Procedures for performing systematic reviews. *Keele, UK, Keele University*, *33*(2004), 1–26.
22. Kotsiantis, S. B., Zaharakis, I., & Pintelas, P. (2007). Supervised machine learning: A review of classification techniques. *Emerging artificial intelligence applications in computer engineering*, *160*(1), 3–24.
23. Krawczyk, B. (2016). Learning from imbalanced data: open challenges and future directions. *Progress in Artificial Intelligence*, *5*(4), 221–232.

24. Larranaga, P., Calvo, B., Santana, R., Bielza, C., Galdiano, J., Inza, I., ... & Santafe, G. (2005).,' A. Perez and V. Robles.." *Machine Learning in Bioinformatics," Briefing in Bioinformatics*, *7*(1), 86–112.
25. Lieber, D., Stolpe, M., Konrad, B., Deuse, J., & Morik, K. (2013). Quality prediction in interlinked manufacturing processes based on supervised & unsupervised machine learning. *Procedia Cirp*, *7*, 193–198.
26. Markham, I. S., Mathieu, R. G., & Wray, B. A. (2000). Kanban setting through artificial intelligence: a comparative study of artificial neural networks and decision trees. *Integrated Manufacturing Systems, 1, 1–33.*
27. McCarthy, J., Minsky, M. L., Rochester, N., & Shannon, C. E. (2006). A proposal for the dartmouth summer research project on artificial intelligence, august 31, 1955. *AI Magazine*, *27*(4), 12–12.
28. Metaxiotis, K., Ergazakis, K., Samouilidis, E., & Psarras, J. (2004). Decision support through knowledge management: the role of the artificial intelligence. *International Journal of Computer Applications in Technology*, *19*(2), 101–106.
29. Mitchell, T., Cohen, W., Hruschka, E., Talukdar, P., Yang, B., Betteridge, J., ... & Krishnamurthy, J. (2018). Never-ending learning. *Communications of the ACM*, *61*(5), 103–115.
30. Oke, S. A. (2008). A literature review on artificial intelligence. *International Journal of Information and Management Sciences*, *19*(4), 535–570.
31. Peng, Y., & Zhang, X. (2007). Guest editorial: Integrative data mining in systems biology: from text to network mining. *Artificial Intelligence in Medicine*, *41*(2), 83–86.
32. Pérez-Ortiz, M., Jiménez-Fernández, S., Gutiérrez, P. A., Alexandre, E., Hervás-Martínez, C., & Salcedo-Sanz, S. (2016). A review of classification problems and algorithms in renewable energy applications. *Energies*, *9*(8), 607.
33. Raynor, W. J., & Shoup, G. (1999). *The International Dictionary of Artificial Intelligence* (Vol.1, pp. 1–33). Global Professional Publishi.
34. Russel, S., & Norvig, P. (2003). Artificial intelligence: A modern approach. *EUA: Prentice Hall*, 1, 170–*178*.
35. Sachs, J. D., Schmidt-Traub, G., Mazzucato, M., Messner, D., Nakicenovic, N., & Rockström, J. (2019). Six transformations to

achieve the sustainable development goals. *Nature Sustainability*, *2*(9), 805–814.

36. Singer, J., Gent, I. P., & Smaill, A. (2000). Backbone fragility and the local search cost peak. *Journal of Artificial Intelligence Research, 12*, 235–270.
37. Stefanuk, V. L., & Zhozhikashvili, A. V. (2002). Productions and rules in artificial intelligence (Vol.1, pp. 1–43). *Kybernetes.*
38. Syam, N., & Sharma, A. (2018). Waiting for a sales renaissance in the fourth industrial revolution: Machine learning and artificial intelligence in sales research and practice. *Industrial Marketing Management*, *69*, 135–146.
39. Tay, D. P., & Ho, D. K. (1992). Artificial intelligence and the mass appraisal of residential apartments. *Journal of Property Valuation and Investment, 1, 1–32.*
40. Vukasin, P. M. (1998). The fundamental problem of an artificial intelligence realization. *Kybernetes*, *27*(1), 71–80.
41. Wang, S., Wang, Y., Du, W., Sun, F., Wang, X., Zhou, C., & Liang, Y. (2007). A multi-approaches-guided genetic algorithm with application to operon prediction. *Artificial Intelligence in Medicine*, *41*(2), 151–159.
42. Wongpinunwatana, N., Ferguson, C., & Bowen, P. (2000). An experimental investigation of the effects of artificial intelligence systems on the training of novice auditors. *Managerial Auditing Journal, 1*, 1–23.
43. Wuest, T., Weimer, D., Irgens, C., & Thoben, K. D. (2016). Machine learning in manufacturing: advantages, challenges, and applications. *Production & Manufacturing Research*, *4*(1), 23–45.
44. Zacharov, I., Arslanov, R., Gunin, M., Stefonishin, D., Bykov, A., Pavlov, S., ... & Fedorov, M. (2019). "Zhores"—Petaflops supercomputer for data-driven modeling, machine learning and artificial intelligence installed in Skolkovo Institute of Science and Technology. *Open Engineering*, *9*(1), 512–520.
45. Zhou, X., Liu, B., Wu, Z., & Feng, Y. (2007). Integrative mining of traditional Chinese medicine literature and MEDLINE for functional gene networks. *Artificial Intelligence in Medicine*, *41*(2), 87–104.

Chapter 6

Materials Discovery and Design Using Machine Learning

CONTENTS

6.1. INTRODUCTION

High-performance material's screening, (QSARs) quantitative structure-activity relationship's modeling and issues regarding material's chemical structures, different biological effects, and other activities of interest are scientifically important as well they are crucial for developing numerous technologically relevant fields (Rajan, 2005; Nantasenamat et al., 2009). Unfortunately, the theoretical characterization and repetitive experimental studies are inefficient and time-consuming, this is because significant progress demands a combination of coincidence and chemical intuition. For example, discovering new materials requires a remarkably long time frame, from initiating the research to its first usage; it takes approximately 10 to 20 years. As Figure 6.1 shows, research for new material comprises seven separate stages which include discovery, development, then the property is optimized and termed as property optimization, then system design and its integration, followed by certification, manufacturing, and finally deployment phase. All these stages might be conducted differently by different scientific or engineering teams at different institutions. Usually, during each phase of the process, experienced and highly skilled teams are involved, few opportunities exist for feedback between initial and final stages, through these opportunities, the process as a whole can be accelerated (Hohenberg & Kohn, 1964; National Science and Technology Council, 2011).

In the material science field, two widely adopted conventional methods are experimental measurement and computational simulation. However, using these methods for accelerating the discovery and design stages is difficult due to the inherent limitations of theoretical foundations and experimental conditions. Speaking generally, experimental measurement usually including property analysis and microstructure, synthetic experiments, property measurement, and a lot more, it is an intuitive and easy method, though it takes place in an inefficient manner (Kohn & Sham, 1965; Binder & Girvin, 1993). Additionally, high requirements are possessed by these approaches in terms of the experimental environment, the researcher's expertise, and the equipment. Alternatively, computational simulation covers a wide range from electronic structure calculations which are based on Monte Carlo techniques, functional theory, the phase-field method, and molecular dynamics to approaches like continuum macroscopic, in such approaches, with the help of computer programs, already existing theories are exploited for analysis (Alder & Wainwright, 1959; Rahman, 1964). It is expected that computation guided material designs will lead towards discovering new materials and will help in reducing the development cost and time

(Chen, 2002; Steinbach, 2009). In contrast to experimental measurement, less time is required by computational simulation, it provides an advantage that it has complete control over the relevant variables. Nonetheless, many challenges regarding computational simulation are present; for example, when studying a new system, previously calculated results cannot be explicitly used; high-performance equipment for computing are required, especially in large computing clusters, on which the programs can run; and it greatly depends on the material's microstructure (Olson, 2000; Boettinger et al., 2002). Close integration between experiment and computation is often required by modern material research, this integration is necessary to yield a basic understanding of the properties and structures of the materials and to also in knowing the fact that how they relate to the processing and synthesis procedures. Using accurate and powerful computational tools, many experiments can be virtually performed, and thus, the time frame can be greatly decreased, in experiments following the traditional model time frame can be 10 to 20 years and by following computational tools, it can be reduced to 18 months (Russell & Stuart, 2003).

Given the advanced necessities associated with understanding the material's fundamental physicochemical properties and accelerating their applications (technological ones), both computational simulation and experimental measurement are incapable to address the emerging new issues. For example, investigating glass's transition temperature utilizing experimental measurements is inefficient and very complicated; since transition takes place at a wide temperature range (Camacho-Zuniga & Ruiz-Trevino, 2003). However, using computer programs does not solve the problem completely as using them the glass's transition temperature cannot be simulated exactly, this is because it depends on many external and internal conditions, such as structure, conformational and constitutive features, and pressure (Yu et al., 2008). Therefore, in the materials science field, many attempts are being made to develop ways so that the shortcomings of these commonly used two methods can be controlled (Belsky et al., 2002; Stanev et al., 2018; French & Bruckman, 2020).

In 2011, when the Materials Genome Initiative (MGI) was launched and the "big data" era was on the way, a huge effort in the community of material science is being made for collecting extensive datasets to provide materials engineers with readymade access to most known material's properties, such as the Materials Commons, the Materials Project (MP), the Harvard Clean Energy Project (HCEP), the Cambridge Structural Databases, the Open Quantum Materials Database (OQMD), the superconducting critical temperatures (SuperCon), the Materials Data Facility, and the Inorganic Crystal Structure Database (ICSD) (Hachmann et al., 2011; Jain et al., 2013). For accelerating materials design and as well its discovery, a generic data sharing and management platform are capable of providing a powerful impetus (Allen, 2002; Kirklin et al., 2015; Liu et al., 2017). Material characterization advanced techniques, with increasing storage capabilities and growing data acquisition rate, raises a great challenge in the field of material science and also for new procedures in analyzing and assessing the collected data (Kusne et al., 2014; Blaiszik et al., 2016). In high-dimensional data, for finding patterns, machine learning is a strong and efficient tool; it provides algorithms, through these algorithms, a computer can grasp the understanding of empirical data using nonlinear or linear relationships for data modeling between the properties of related factors and materials (Murphy, 2012; Raccuglia et al., 2016). Over the past few years, big data methods and machine learning techniques have resolved many issues of modeling the relations between complex physical factors and materials properties (Ghiringhelli et al., 2015; Mueller et al., 2016). In particular, machine learning's successful applications are considered as an innovative model in the field of materials development, for instance, screening performed for molecular organic diodes (light-emitting ones) and assisting in materials discovery, in which new hypotheses about the requirements for successful product formation are revealed by the machine learning models (Kalidindi et al., 2016; Ward & Wolverton, 2017).

Over the last two decades, the computational activities associated with materials science have abruptly shifted from purely computational studies and technique development towards the design and discovery of new materials, totally in the guidance of the machine learning, data mining, and computational results or by collaboration between experimental validation and computational predictions (Hautier et al., 2012; Chen et al., 2015).

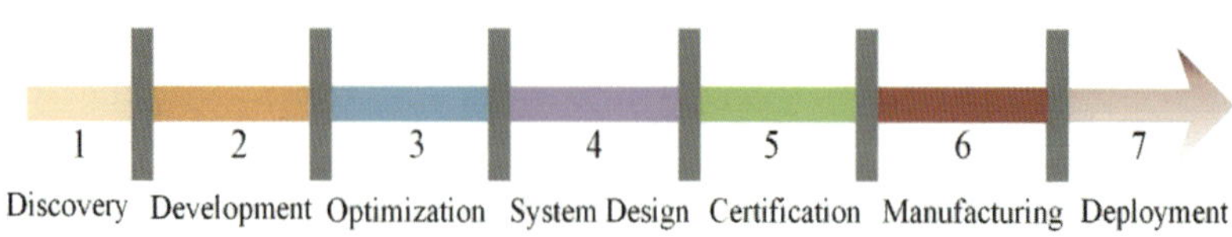

Figure 6.1. New materials finding process by using traditional methods.

[Source: https://obamawhitehouse.archives.gov/sites/default/files/microsites/ostp/materials_genome_initiative-final.pdf.]

The benefits of strategies regarding modern materials research depend upon their ability in finding a harmony between a low error rate and reasonable experimental requirements to utilize the available data effectively and for making the materials research process faster (Hill et al., 2016; Jain et al., 2016). Huge efforts are being made for developing methods that are more suitable and can combine techniques of intelligent data analysis along with the traditional experimental methods; this will help in reducing the error rate and in improving the experimental efficiency. For example, for synthesis guidance of new inorganic materials, a novel integrated method was proposed by Sumpter et al. (2015), this idea incorporates approaches of big data in scattering and imaging combined with first principles that are scalable. The techniques associated with deep data and big data, helps in linking the theoretical predictions of first-principles with high-veracity scattering and imaging data from materials having microscopic freedom degrees, thereby accelerating rational materials synthesis and design. For novel ternary compounds, as reported by Meredig et al. (2014) and Haughtier et al. (2010), an iterative combination of first-principles calculations and machine learning methods can greatly help in accelerating the discovery process (Agrawal & Choudhary, 2016; Gómez-Bombarelli et al., 2016).

6.2. MACHINE LEARNING METHODS DESCRIPTION IN MATERIALS SCIENCE

For artificial intelligence, machine learning emerged out of the hunt (Rosenblatt, 1958). Attempts were made in the 1950s with aim of approaching the problems that machine faced while acquiring the knowledge by using different symbolic methods, and then later, connection principle-based methods, such as perceptron and neural networks, were widely studied (Vahed & Omlin, 1999; Vapnik, 2013). Afterward, various methods based on statistical learning theory (SLT), which include decision trees (DTs) and support vector machines (SVMs) were proposed. Nowadays various new machine methods like deep learning which are used for analyzing big data have inspired both industry and academia (De'ath & Fabricius, 2000; Ustun et al., 2017). Machine learning is based on the automation of the analytical building model. It uses the algorithms that learn from data iteratively, in this way it permits computers to search and detect hidden insights without being directly programmed (Larranaga et al., 2006; Joze & Drew, 2010). Good applicability is shown by machine learning in regression, classification, and other high-dimensional data-related tasks. Aimed at gaining insight and extracting knowledge from huge databases, machine learning uses the knowledge of previous computations and learns from them to produce repeatable, reliable results and decisions and hence has played a key role in numerous fields, such as natural language processing (NLP), bioinformatics, image recognition, information security, and speech recognition (Rao & Mukherjee, 1996; Olsson, 2009; Eminagaoglu & Eren, 2010). Machine learning algorithms powers numerous daily based activities including credit scoring, text-based sentiment analysis, next-best offers, web searches, and fraud detection. In material science, machine learning's pioneering applications can be traced back to the era of 1990s, at that time, it's methods such as artificial neural networks (ANNs) and symbol methods were in practice for predicting the compressive and tensile strengths and the corrosion behavior of the matrix/ fiber interfaces in composites of ceramic-matrix (Reich & Travitzky, 1995; Chonghe et al., 1996). Afterward, in materials science, machine learning methods have been frequently used in addressing several topics such as material property prediction and new materials discovery (Mitchell, 1999).

6.2.1. Machine Learning Paradigms in Materials Science

Classically, machine learning is defined as <P, T, E>, where P denotes performance, T denotes task, and E denotes experience. Mainly, it is interpreted in a way that a computer program learns from E (experience) referring to a class of T (tasks) and P (performance), with experience E, performance P improves on a task T (Peck et al., 2015). Generally, in materials science, a machine learning system is used when machine learning is being used to solve or address a given problem. The machine learning system's general paradigm is given below:

$$\text{Model} = \text{Goal} + \text{Sample} + \text{Algorithm} \qquad (1)$$

Here, the given problem is represented by the ultimate Goal that is mostly expressed as an objective function. The Sample is referring to the population's subset that has been selected to study (Wu, 2013). Generally, data preprocessing which includes feature engineering and data cleaning is employed for extracting the sample from the original data. Feature engineering, together with feature construction, feature extraction, feature learning, and feature selection refers to a process in which domain knowledge about the data is used for creating features that allow the algorithms (machine learning) to function. To the machine learning applications, feature engineering is fundamental and is both expensive and difficult. Data cleaning is a process in which inaccurate, incomplete, irrelevant, and incorrect parts of the data are identified and then this coarse or dirty data is replaced, modified, or deleted (Provost & Kolluri, 1999). Both the algorithms, the model optimization, and the machine learning algorithm, is a step-by-step, self-contained set of operations that are to be performed (Fujimura et al., 2013). ANN, DT, and SVM algorithms are the machine learning algorithms that are most commonly used. Particle swarm optimization (PSO), genetic algorithms (GAs), and simulated annealing algorithms (SAAs) are mainly included in the model optimization algorithms. The Model refers to those algorithms that are learned from the Sample and it describes the system in mathematical terms.

6.2.2. Major Phases of Machine Learning Usage in Materials Science

As Figure 6.2 shows, the machine learning system's construction is divided into three phases: first comes sample, then construction, and finally model building and its evaluation.

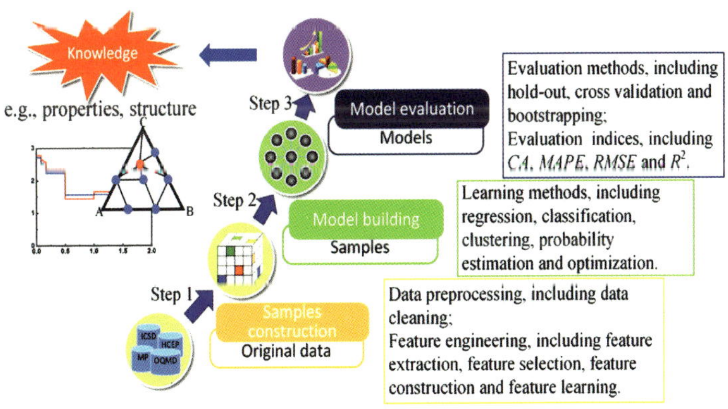

Figure 6.2. The general machine learning process in materials science.

[Source:https://www.sciencedirect.com/science/article/pii/S2352847817300515.]

6.2.2.1. Sample construction is the first phase

The original data in materials science are collected from experimental measurements and computational simulations. These data are usually noisy, inconsistent, and incomplete, and hence, before constructing a sample, data cleaning has to be performed on the original data. Moreover, an obtained sample can be affected by several conditional factors, among which some may not relate to the decision attributes, e.g., in researches of predicting the Li ionic conductivity, it is expected that various external and internal factors may affect the conductivity, but the most relevant factors such as transition temperature, ion diffusivity, experimental temperature, and average volume, are appealing in comparative experiments (Guyon & Elisseeff, 2003). Thus, it is necessary to use and choose a feature selection method that can properly determine the attribute's subset, so that it can be utilized in the final simulation (Hutchinson, 2009).

6.2.2.2. Model building is the second phase

The model building phase is a black box, which plays its essential part in linking the input and output data. It uses a particular set of linear or nonlinear functions. In materials science, for typical research, usually, complex relationships exist between the target attributes and the conditional factors, by using traditional methods, such relationships are difficult to handle. A medium to use target function examples is provided by machine learning through which the coefficients can be found by using a certain mapping function, e.g., in glass transition temperature T_g predicting experiment T_g (Efron, B., & Tibshirani, 1994), there are great difficulties in finding a formula that precisely describes the relation among T_g and other relevant factors of average molecular polarizability, rigidness, net charge, and chain mobility: however, by using a machine learning method, the relationships between decision attributes and conditional factors can be modeled. Here, the "core" algorithms lie and it is the point where machine learning has a vital role. The knowledge acquired by using machine learning is stored. It is stored in a readily usable format and, successively, can be utilized for materials design and discovery.

6.2.2.3. Model Evaluation is the Last Phase

A good performance on existing data and as well on unseen data should be achieved by a data-driven model. In general, the model's generalization errors can be evaluated by using calculation-based tests, further, this result can help in selecting the best one. Thereby, we require testing data so that the model's discriminative capabilities can be tested on a new dataset; after testing, the error achieved can be used as an approximation for generalization error. When only one dataset, containing m samples is present D= $\{(x_1, y_1,),$

(x_2, y_2), …, $(x_m, y_m)\}$, the dataset can be partitioned into two datasets; from which one can be used as a testing dataset T and other as training dataset S by using various evaluation methods, some of them are shown in Table 6.1.

The dataset D, in the hold-out method, is partitioned into a testing dataset T and a training dataset S in a way that $S \cap T = \emptyset$ and $D = S \cup T$. Stratified sampling is a category-preserving sampling method that keeps the data distribution consistent to the highest possible extent, it also avoids the introduction of deviations (additional ones). As no perfect solution exists that can decide the relative proportions of T and S, in D, 2/3e4/5 of the samples are considered as S and the remaining are considered as T.

The dataset D, in the cross-validation method, is partitioned into same sized k subsets that are mutually exclusive, $D = D_1 \cup D_2 \cup \ldots \cup D_k$, $D_i \cap D_j = \emptyset (i \neq j)$, here "stratified sampling" generates each D_i. Then, the k-1 union set of the subsets is considered as S (training dataset), and the remaining one is considered as T (testing dataset). Leave-one-out cross-validation (LOOCV) is a special case that takes place when in D (original dataset, number of samples m is equal to k, random sample partitioning does not affect it. For every single subset of data, this process is repeated, and thus, k numbers of experiments are conducted in total. Therefore, for large datasets, the cross-validation method is unsuitable and may take a lot of time.

The method of bootstrapping is based on bootstrapping sampling, it involves a sample from the dataset, copies it to a new dataset D0 repeatedly and randomly until the new sample contains m samples (Bishop, 2006). Then, this new dataset is considered as the training dataset, and for the testing dataset, D\D0 is used. The number of the training sample is maintained equal to the original dataset's size, this method plays an effective role when we have small data volume and are facing difficulty in partitioning the testing/ training data properly. However, this method potentially introduces estimation bias as it changes the original dataset's distribution.

An algorithm when evaluated undergoes a test to see its performance. The accuracy that how well is the prediction of the model is examined by comparing the corresponding predicted values with the experimental values. The standard of evaluation depends on the problem type encountered. In classification problems, to evaluate the models being used, we use classification accuracy (CA):

$$CA = S/N \quad (2)$$

where S denotes the number of correctly classified samples and N denotes the total number of samples.

Table 6.1. The Comparison between Different Evaluation Methods

Method	**Advantages**	**Disadvantages**	**Applicable situation**
Hold-out	Low computational complexity.	It is difficult to determine the proper proportions (relative) of testing and training data; Compared to the original dataset, the training data has a small volume.	The volume of data is sufficient.
Cross-validation LOOCV	Changing the training data volume does not greatly influence this model.	It has high computational complexity, especially on large datasets; Compared to the original dataset, the training data has a small volume.	The volume of data is sufficient. The volume of data is small, and the testing and training data can be effectively partitioned.
Bootstrapping	Effective partitioning of both, the testing and training data.	The training data has a different distribution from the original dataset.	The volume of data is small, and it is difficult to partition the testing and training data.

As expression (3)–(5) shows, the correlation coefficient (R^2), the root mean square error (RMSE), and the mean absolute percent error (MAPE), these all are used for evaluating the models applied in solving the regression problems.

$$MAPE = \frac{1}{n}\sum_{i=1}^{n}\frac{|y_i' - y_i|}{y_i} \quad (3)$$

$$RMSE = \sqrt{\frac{1}{n}\sum_{i=1}^{n}(y_i' - y_i)^2} \quad (4)$$

$$R^2 = \frac{\left[\sum_{i=1}^{n}(y_i - \bar{y})(y_i' - \bar{y}')\right]^2}{\sum_{i=1}^{n}(y_i - \bar{y})^2 \cdot \sum_{i=1}^{n}(y_i' - \bar{y}')^2} \quad (5)$$

where y_i represents an original value and $y_i^{'}$ represents the corresponding predicted value, and y_0 is the averages of the predicted values and y of the original value.

Moreover, to evaluate the accuracy of the classification model, other indices include the ranking loss, the coverage error, the Jaccard similarity coefficient, the label ranking precision (average), the Hamming distance, the confusion matrix, the hinge loss, Cohen's kappa, the logistic regression loss, and the receiver operating characteristic curve (ROC), the recall, and the precision. The two additionally used indices for regression problems are the coefficient of determination and the explained variance. In clustering problems, for model evaluation, the commonly used indices are the silhouette coefficient, mutual information (MI), and the Rand index.

6.2.3. Machine Learning Algorithms Commonly Used in Materials Science

While constructing a machine learning system, it is important to select a machine learning algorithm that is appropriate, as it can greatly affect the generalization ability and prediction accuracy (Pei et al., 2013). No such suitable algorithm exists that can be used for all problems, and thus, there is a different scope of application for each algorithm. As Figure 6.3 shows, in materials science, we can divide the machine learning algorithms that are most commonly used into four groups: regression, probability estimation, clustering, and classification. The algorithms of probability estimation are used for the discovery of new material, whereas for material property prediction, clustering, classification, and regression algorithms are used on the micro and macro-levels (Fang et al., 2008; Paszkowicz, 2013). Besides, the methods of machine learning are usually combined with several intelligent optimization algorithms, for example, PSO, SAAs, or GAs algorithms, these are for the optimization of the model parameters (Zhang et al., 2008; Mohn & Kob, 2009). Moreover, we can use these optimization algorithms for performing other difficult optimization tasks, for instance, materials properties and spatial configurations optimization (Ning et al., 2012; Isayev et al., 2017).

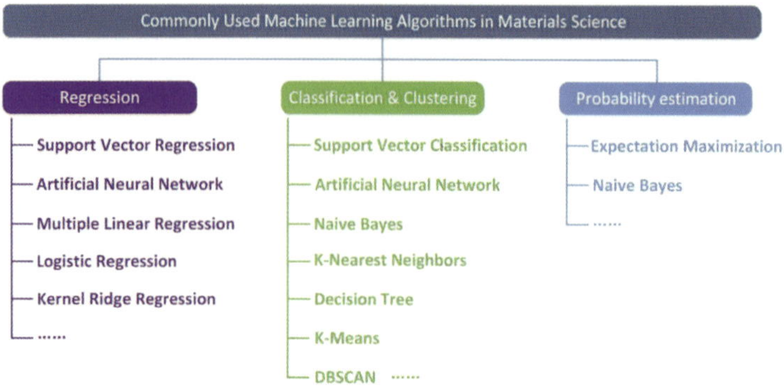

Figure 6.3. Machine learning algorithms that are commonly used in materials science.

[Source:https://www.sciencedirect.com/science/article/pii/S2352847817300515.]

6.2.4. Overview of the Machine Learning Applications in Materials Science

In materials science, we widely use machine learning. Machine learning demonstrates superiority in prediction accuracy and also in time efficiency. As Figure 6.4 shows, we can divide the machine learning applications in materials design and discovery into three main categories: new materials discovery, material property prediction, and several other purposes. Methods of regression analysis are usually used in the researches regarding material property prediction, and we can predict both microscopic and macroscopic properties. In discovering new materials the basic purpose behind the application of machine learning is utilizing a probabilistic model to present structures and components in different combinations and ultimately to choose a material from the candidate set that performs better in terms of validation based on density functional theory (DFT). Moreover, in materials science, we use machine learning for other purposes such as density function approximation and process optimization.

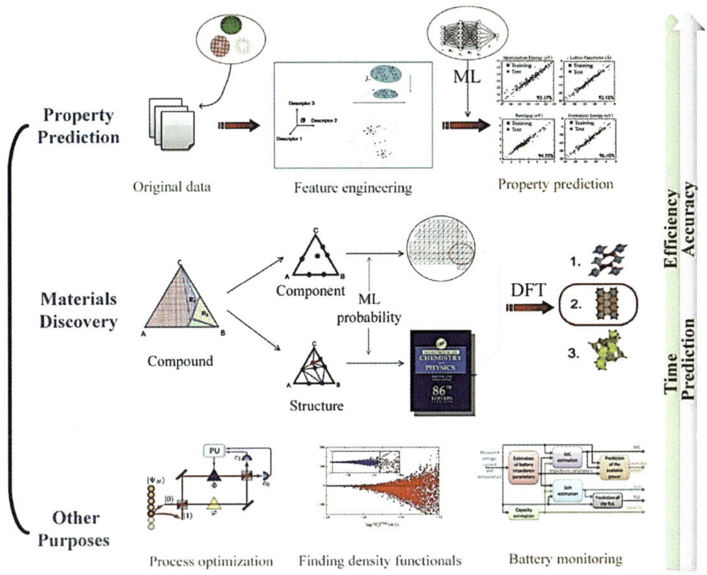

Figure 6.4. Overview of the machine learning applications in materials science.

[Source:https://www.sciencedirect.com/science/article/pii/S2352847817300515.]

6.3. THE MACHINE LEARNING APPLICATIONS USED IN MATERIAL PROPERTY PREDICTION

We can describe the different properties of a material, for instance, molecular atomization energy, ionic conductivity, the lattice constant, melting point, glass transition temperature, and hardness at either the microscopic or macroscopic level. Mostly, two methods are used to study the properties of the material: experimental measurement and computational simulation, both these methods involve experimental setup and complicated operations. Hence, building such computational simulations that can completely capture all the logically complicated relationships between the material's properties and related factor is quite difficult and complicated, especially when few such relations might be unknown. Furthermore, experiments performed for measuring the compound's properties usually occur in the subsequent materials selection stages. In consequence, if satisfactory results are not obtained, the massive amount of experimental resources and time invested turns to be a waste. Additionally, in numerous cases, studying the material's properties is almost impossible or very difficult, irrespective of the amount of experimental and computational efforts made. Therefore, it is necessary to develop high-performance and intelligent prediction models on urgent bases so that the material properties can be correctly predicted at a low computational and temporal cost. The main concern of machine learning is the study and construction of algorithms that can acquire the knowledge of patterns from data. For predicting material property, the basic aim of using these methods (machine learning) is the analysis and mapping of relationships, mostly non-linear ones, between the material properties and related factors, that should be done by using the extracted knowledge from previously existing empirical data.

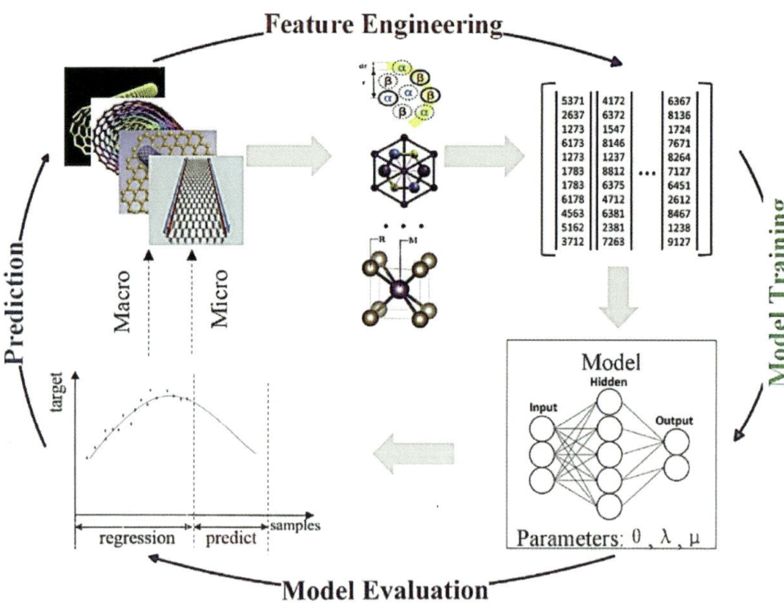

Figure 6.5. The basic framework for machine learning applications in material property prediction.

[Source:https://www.sciencedirect.com/science/article/pii/S2352847817300515.]

In Figure 6.5, machine learning application's fundamental framework in material property prediction is shown. Firstly, feature engineering or hand tuning (including the selection and extraction of feature) is conducted for identifying the conditional attributes correlated to property prediction. Secondly, through model training, a mapping relationship is found between the decision attributes and conditional factors. Lastly, the trained model is used for predicting the materials property. For example, Property-Labeled Materials Fragments (PLMF) named calculation tool was proposed by Isayev et al. (2017), this tool is used for building such models that can predict the inorganic crystal's properties. In PLMF, highly correlated and low-variance features are initially filtered out, so that a feature vector can be obtained. The technique of the gradient boosting decision tree (GBDT) will be used to classify a novel candidate material as either an insulator or a metal, and if the material turns out to be an insulator, the bandgap energy is predicted. Regardless of the metal or insulator classification of material, thermomechanical properties (thermal expansion coefficient, the capacity of heat at constant volume, the capacity of heat at constant pressure, shear modulus, bulk modulus, and Debye temperature) are then predicted. Before training the model, the dataset is partitioned by using fivefold cross-validation. The MAE, RMSE, R^2, and ROC curves are used for evaluating the trained model's prediction accuracy. In material property prediction, depending on the analysis scale, we can divide the machine learning applications into two categories. The first category is named macroscopic performance prediction and the second category is named microscopic property prediction.

6.3.1. Macroscopic Performance Prediction

Research on material's macroscopic performance centers around the structure-activity relationship, present between the material's macroscopic properties (e.g., physical and mechanical) and its microstructure (Friedman, 2001; Carrera et al., 2008). Due to their good performance in classification and regression related problems, machine learning approaches that involve SVM and ANN algorithms combined with optimization algorithms are widely used in the macroscopic performance prediction studies (Bertinetto et al., 2009).

ANNs are models in machine learning that are based on the idea of biological neural networks (that is animal's central nervous system (particularly the brain)), these models are used to approximate or estimate functions that are generally known and can rely on a greater number of inputs. This approach (ANN) is a statistical (nonlinear) analysis technique; it has strong adaptive and self-learning capabilities (Wang, 2003; Guo et al., 2005). Material behaviors such as tensile and temperature responses, wastage, elongation, compressive, and corrosion properties are usually predicted by Backpropagation ANNs (BP-ANNs) (Topcu & Sarıdemir, 2007; Altun et al.,

2008). A linear regression a BP-ANN model was employed by Chen et al. (2008) for predicting polymer's glass transition temperatures and the average prediction error came out to be much smaller for the former approach (17 K) than the error predicted by the latter approach (30 K). As there is no requirement of physical background knowledge, dramatic benefits can be provided by BP-ANNs to the industry, as it can solve problems with good generalization ability and acceptable prediction errors. However, typically their convergence rate is low and they even can fall into local minima. On the other hand, another type of ANN, named RBF-ANNs in which the radial basis function is combined with the ANN concept, can overcome the local minima problem and this approach has a high rate of convergence. It was reported by Gajewski and Sadowski (2014) that using RBFANN for investigating crack propagation in a pavement structure that is bituminous layered, this cracking increases bituminous layer B2 thickness. Furthermore, ANN modeling is now used in other applications too, i.e., predicting melting points, the viscosity of biofuel compounds, density, diffusion barriers, and excited-state energies (Saldana et al., 2012; Salahinejad et al., 2013; Häse et al., 2016). ANN's biggest advantage is its capability to be utilized as a function (arbitrary) approximation mechanism that depends and learns from observed data. ANN model utilization revolves around the fact that a small amount of prior knowledge regarding the target material is needed and from empirical data, we can learn the rules that govern property variations (Raj & Daniel, 2008; Sivasankaran et al., 2009). ANNs are commonly criticized because of the reason that they require a highly diverse training dataset, along with abundant representative examples which predict the property for capturing the underlying structure to such an extent that the results can be used and generalized to new cases. Moreover, neural networks have an inherited deficiency that the acquired knowledge is concealed in numerous connections, leading to poor comprehensibility, for instance, poor explainability and poor transparency of knowledge (Scott et al., 2007; Wu et al., 2017).

The SVM technique is a supervised learning method based on SLT; it is used for recognizing patterns and analyzing data that are used for regression and classification (Cortes & Vapnik, 1995; Cavaliere, 2007).In contrast to ANNs models, SVM models are better and more appropriate for applications having small samples, they are capable to overcome the issues of "overlearning" and "the curse of dimensionality." Numerous unique advantages for solving high-dimensional and nonlinear problems are exhibited by them. A novel hybrid methodology was proposed by Fang et al.

(2008), it combines support vector regression (SVR) and GAs, which makes it competent in forecasting the metallic material's atmospheric corrosion behaviors, i.e., steel and zinc. When compared to other methods, results prove that a better predictive capability is provided by this model, and hence, it is considered as a promising alternative forecasting method, which forecasts the corrosion behaviors of steel and zinc. A forecasting method named as Feature Selection Two-Stage SVR (FSTS-SVR) was proposed by Liu et al. (2019) to overcome a single SVR model's lower performance that predicts the As-Se glass transition's temperature. FSTS-SVR is used to investigate the turning point depending on the structural analysis and for each stage, a prediction model was constructed. The experimental results show that in contrast to other methods (that are SVR based and involves a turning point), the FSTS-SVR method's prediction accuracy is higher (Gharagheizi et al., 2012; Alzghoul et al., 2014). An SVM approach proposed by Chen et al. (2009) was used for estimating the exposure temperatures of concrete structures damaged by fire. Based on the results, concrete's ultrasonic pulse velocity was identified to be the most effective parameter that can improve estimation accuracy. Moreover, we can employ SVM models to predict glass transition temperatures, ionic conductivities, and several other behaviors of functional materials (Xu et al., 2009; Pei et al., 2013).

6.3.2. Microscopic Property Prediction

Microscopic properties determine the material's macroscopic performance, including its structural (which includes lattice constant) and atomic characteristics. As far as our knowledge, the machine learning applications predicting microscopic property concentrate on various aspects, including the electron affinity, the lattice constant, molecular atomization energy, and band energy.

In the design of buffer materials or buffer materials for the semiconductor epitaxy, the major concern is the mismatch among the layers comprising different materials. Therefore, predictions regarding lattice constant may help in accelerating the lattice constants for all combinations of elements. Regarding this, a reliable and fast solution to predict the lattice constants of numerous unknown compounds is raising as a fundamental requirement in developing high technology. This prediction problem can be solved by applying machine learning because of its better capability in regression analysis (Helma et al., 2004; Yu & Cheng, 2007). As Table 6.2 shows, several studies regarding the usage of machine learning for predicting lattice constant have been done. These studies were done for perovskite crystal materials. Regression methods of machine learning such as logistic regression (LR), ANN, and SVR techniques have been used in these studies. By using a sample set containing 157 known $GdFeO_3$-type ABO_3 perovskites and ANN and LR techniques, they correlated lattice constants (along with their constituent elemental properties) in $GdFeO_3$-type ABO_3 perovskites (Chonghe et al., 2003).

The first obtained LR models used two elemental ionic radii. Similarly, ANN models were created focusing on five basic elemental properties. Later, the valence of ion A, the ionic radii, and the cations A and B electro-negativities were considered too, aiming that it will improve the model's predictive capabilities. The results suggested that the percentages of absolute difference (PADs) of 0.93 for constant a, 0.82 for constant b, and 0.77 for constant c (0.35, 0.34, and 0.44), respectively, were achieved by the LR (ANN) method. Compared to the LR method, the ANN had a better accuracy as it achieved an error within 2%. Moreover, the prediction accuracies on the testing and training data of the ANN method were compared and according to the results, the prediction accuracy was high on the training data, while accuracy for the testing data was unsatisfactory. SVR was employed for predicting orthorhombic ABO_3 perovskite compound's lattice constants; it was employed by Javed et al. (2007) for overcoming the weak generalization ability of ANNs. The result was satisfactory as the performance of the SVR model on both the testing and training datasets was better than the ANN. The average PAD values for all lattice constants were less than 1%.

Additionally, compared to the ANN model, less testing and training time was required by the SVR model. ANN, generalized regression neural network (GRNN), and SVR methods were used in Majid et al. (2010) for predicting the perovskite compound's lattice constants. The mean PAD value of the ANN model was 0.96%, GRNN was 0.54%, and SVR was 0.43%, which indicates that the best prediction performance was shown by the SVR method. Several machine learning methods were used by Majid et al. (2011), these methods include multiple linear regressions (MLR), random forests (RF), a GRNN, and SVR and to predict the lattice constant, the only conditional attribute chosen was the ionic radius. According to results, the prediction accuracies yielded by MLR was 0.51, by RF was 0.41, by a GRNN was 0.70, and by SVR was 0.26, all of these are much higher than the value yielded by well-known SPuDS software; this software is abundantly used in crystallography (Lufaso & Woodward, 2001). The prediction model's performances were compared with each other. Relating the predicted and experimental values for the chosen dataset, the performance curves are shown in Figure 6.6. We can see from the figure that the experimental values which are represented by R^2 for all the used methods were greater than 0.95, which indicates the good performance in lattice constant prediction of all the four machine learning methods. Moreover, when there was a small training sample, in contrast to GRNN, RF, and ANN methods, the SVR method turned out to possess a higher generalization ability.

Table 6.2. Machine Learning Applications for Predicting the Lattice Constant

Conditional Attributes	**Compounds**	**ML Methods**	**PAD**		
			A	**b**	**c**
ionic radii (r), electro-negativities (x), valence (z).	$GdFeO_3$-type ABO_3 perovskites	LR/ANN	0.93/0.35	0.82/0.44	0.77/0.34
electro-negativities (x),ionic radii (r), valence (z).	ABO_3 perovskites	SVR	0.52	0.54	0.58
ionic radii (r), electro-negativities (x), oxidation state (z).	perovskites	GRNN/ SVR/ANN	average of a is0.43, of b is0.54, and of c is 0.96		
ionic radii (r), electro-negativities (x), oxidation state (z).	cubic perovskites	RF/GRNN/ SVR/MLR	average of a is 0.26, of b is0.41, and of c is 0.51		

Five machine learning methods were outlined by Hansen et al. (2013), which include the ANN method, the method of K-nearest-neighbor (KNN), SVR, kernel ridge regression (KRR), and linear ridge regression (LRR) to investigate the molecular representation's influence on the methods' performance. As a result, quantum-chemical computations were allowed by all five methods, these computations were performed in milliseconds rather than days (or hours) when ab initio calculations were used and the lowest prediction error was achieved by the kernel ridge regression method, which was 3kcal/mol, for the atomization energies. By using the SVR, RF, and Pearson coefficient, a two-phase hybrid system was constructed by Liu et al. (2015), this system includes regression and feature extraction. On 2500 volume elements (micro-scale), experimental results show that elastic localization linkages can be built by this system with testing (training) MASE (mean absolute strain error) of 7.17% (13.02%).

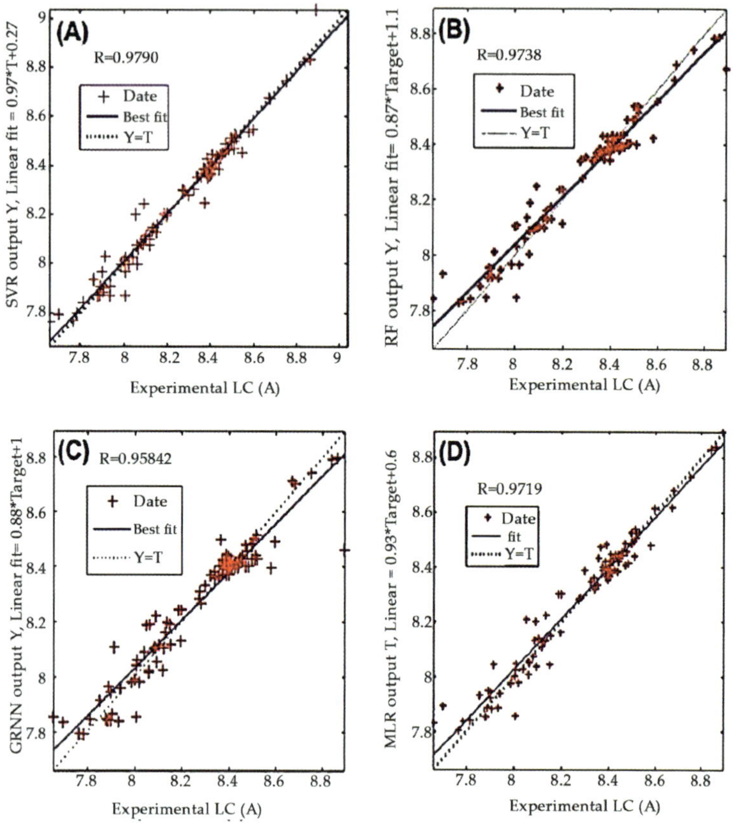

Figure 6.6. (A-D) Linear correlation between the predicted and experimental values is shown for each model of machine learning, where the original data is represented by the red points, the blue curvy line fits the results, and for reference, there is the dotted line.

[Source:https://www.sciencedirect.com/science/article/pii/S2352847817300515.]

Recently, a multi-agent data-driven model for prediction was constructed by these authors, they introduced principal component analysis (PCA) and K-means clustering for improving a single-agent system's performance (Liu et al., 2017). According to the results, for both testing and training, better MASE values can be achieved by the multi-agent model. Three key strategies were applied by Ward et al. (2016) for improving prediction accuracy and time efficiency of machine learning methods used to predict the glass-forming ability and bandgap energies of inorganic materials. These applied strategies aim to construct a general-purpose framework of machine learning with improved accuracy and efficiency. Firstly, an attribute set (general-purpose) comprising 145 elemental properties was summarized, which can capture the decision properties effectively. Secondly, for overcoming the disadvantages of every individual method, they used an ensemble learning method.

Thirdly, a partitioning strategy was utilized by grouping the dataset elements into chemically similar subsets, and then, a separate model on these subsets was trained. By employing statistical or machine learning methods that are trained on an idea of chemical similarity and based on quantum mechanical computations, it was reported by Pilania et al. (2013) that a diverse set of material system's microscopic properties can be predicted accurately and efficiently (as in Figure 6.7). It is the first example of predicting polymer material's microscopic properties using an approach that is based on a distance-measurement. This proposed methodology for predicting microscopic property has four steps. Firstly, within a class, material motifs are reduced to fingerprint (numerical) vectors. After this, a suitable measure of chemical distance or chemical similarity is used for the evaluation of each material. Finally, in this case, a learning scheme KKR is employed for mapping the relationship between the properties and distances. Fingerprints based on either electronic charge density distribution or chemo-structural (configurational and compositional) information can be used to make property predictions that are accurate and ultra-fast. In material property prediction, machine learning methods are applied for lithium-ion batteries.

The method of the DBSCAN (Density-Based Spatial Clustering of Applications with Noise) was combined by Chen et al. (2017) for elucidating the diffusion mechanism of Li in the $Li_7La_3Zr_2O_{12}$ crystal lattice. Later, it was proved that we can use DBSCAN for recognizing the lattice sites, for determining the type of site, and for identifying the Li hopping events. In lithium-ion batteries, for cathode volume changes, QSAR formulations were constructed by Wang et al. (2017), in them, the method of the partial least squares (PLS) was used on the data that was gathered from ab initio calculations. The analysis of VIP (variable importance in projection) indicated that during delithiation, the change in volume of a cathode was strongly affected by the X octahedron descriptors and the radius of the X⁴þ ion. Furthermore, ANNs have been efficiently for predicting formation energies and vacancy migration, potential energies, electron affinities, and vacancy migration energies, thereby promoting a better and deeper physical understanding of chemical system's microscopic properties (Castin et al., 2014; Schütt et al., 2014; Von Lilienfeld et al., 2015).

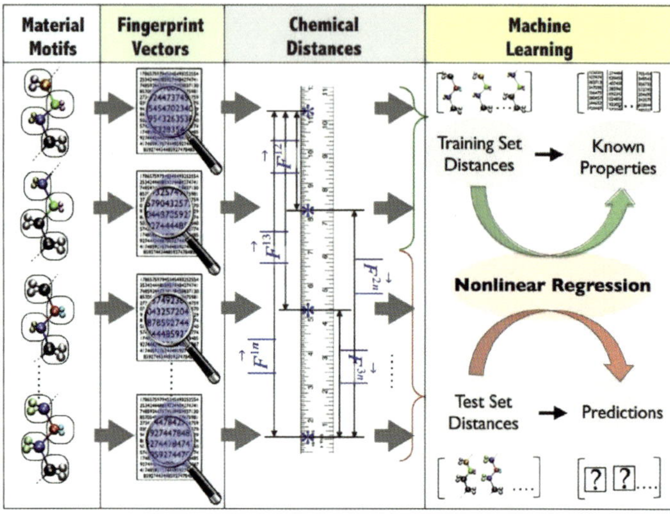

Figure 6.7. A machine learning methodology used for distance measurements for predicting the microscopic properties of materials.

[Source:https://www.sciencedirect.com/science/article/pii/S2352847817300515.]

As the above analysis show, in contrast to traditional methods, machine learning methods of material property prediction (micro and macro) exhibit higher robustness and accuracy, although the performance solely firmly depends on the sample construction and the selected method. It is worthy to mention that there is not a single method (machine learning) that can be used to achieve good outcomes for all applications. Hence, there is always a need to compare various machine learning methods for selecting the best solutions. Feature engineering and data cleaning are elementary steps of sample construction, they help in improving the accuracy of prediction of a trained model as they optimize the features. For verifying the feature selection's importance, experiments were conducted by us in which we used five different material property datasets that were using MLFFS (Multi-Layer Filtering Feature Selection). MLFFS was proposed in Gu et al. (2018), it employs RF, the Pearson coefficient, PCA, and variance filtering to automatically and sequentially delete irrelevant and sparse redundant features. Table 6.3 lists the results. As it shows that in all five datasets, the numbers of features decrease after feature selection, while an increase in terms of R^2, MAPE, and RMSE values is observed in the model prediction accuracies. Especially, in Dataset 4 and 5, the numbers of features are reduced to nearly half, with higher R^2 values and lower MAPE and RMSE values. Therefore, we can conclude that the accuracy of property prediction of a material can be improved by feature selection, and for materials science researchers, MLFFS might become an efficient tool.

6.4. THE USE OF MACHINE LEARNING APPLICATIONS IN THE DISCOVERY OF NEW MATERIALS

In materials science, a permanent topic is to find new materials that can show better performance. Currently, for the discovery of new materials, computational and experimental screenings involve structure transformation and element replacement. However, both or one of the structural search space or compositional search space, tend to be strongly constrained. A massive amount of experimentation or computation is required by both screening methods and it results in an "exhaustive search," consuming resources and time. While considering the benefits of machine learning, an adaptive method that combines computational simulation and machine learning is proposed to perform screening and to evaluate the new materials "in silico" for providing suggestions for better and new materials.

Fig. 6.8 shows the general machine learning process for discovering new materials. For this purpose, the machine learning system involves two parts, i.e., a prediction system and a learning system. The model obtained from the learning system is applied by the prediction system for structure and component prediction. On the other hand, the operations of feature selection, data cleaning, and model testing and training are applied by the prediction system. Frequently, a suggestion-and-test approach predicts new materials: the prediction system selects candidate structures through structure recommendation and composition recommendation, and for comparing their relative stability, DFT calculations are used.

Presently, for finding new materials, numerous machine learning methods with good performance are being used (as Table 6.4 shows). We can divide them into methods focused on composition prediction and crystal structure prediction, the following subsections will discuss it in detail. However, an exception is represented by Farrusseng et al. (2009), in which GA and ANN approaches are integrated but failed to accelerate the discovery of new materials since there are no well-known relevant descriptors: nonetheless, for the discovery of new materials, we can regard this work as an implementation of a new idea.

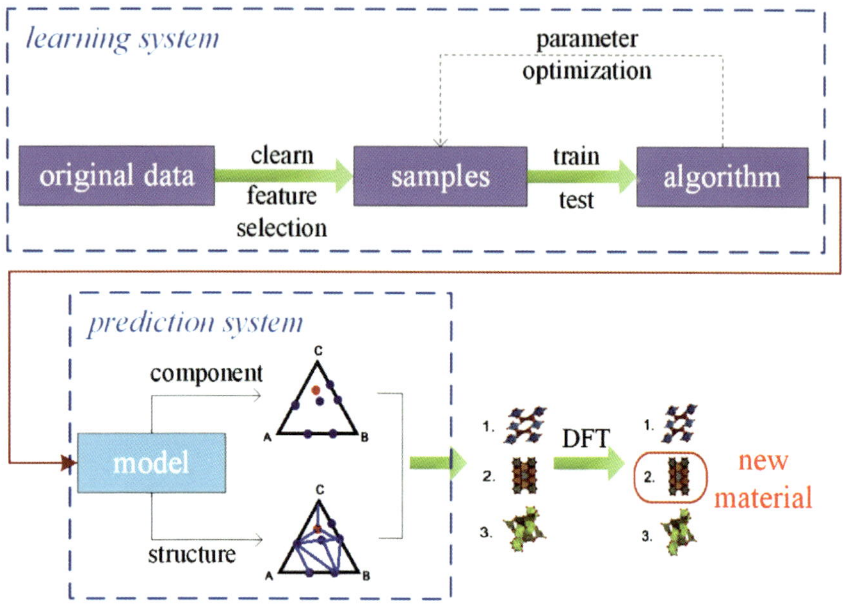

Figure 6.8. The general machine learning process that is used for discovering new materials.

[Source:https://www.sciencedirect.com/science/article/pii/S2352847817300515.]

6.4.1. Crystal Structure Prediction

The characterization and prediction of a material's crystal structure constitute a major problem in forming the basis for materials design. We can avoid unnecessary structure-related experiments by using crystal structure prediction; this will help in reducing the consumption of computing resources and DFT calculation. It is even more difficult to predict crystal structures while following the chemical-based reactions, as for the entire reaction, it demands a precise potential energy surface. Even for the simplest of crystallization, it is fundamentally difficult to predict crystal structure using the first principle technique; this is due to its requirement of considering combinatorically huge component arrangement sets using the high-level methods of quantum chemistry (Beran, 2015). Machine learning focuses on the empirical rules extracted by algorithms from the experimental information, and this approach is catching the attention of everyone.

Table 6.3. Predicted Results of Five Datasets While Using MLFFS for Feature Selection

Dataset	Original Features	Original Prediction Result			Selected Features	Final Prediction Result		
		R2	MAPE	RMSE		R2	MAPE	RMSE
1	6	0.96	0.26	0.59	4	0.97	0.26	0.58
2	6	0.93	0.01	0.08	5	0.90	0.01	0.07
3	5	0.96	0.09	8.69	4	0.96	0.07	7.24
4	47	0.97	0.03	28.36	25	0.98	0.03	25.16
5	18	0.91	0.27	0.64	9	0.97	0.15	0.36

Before the 1980s, no one paid attention to crystal structure prediction researches. In materials science, the need for crystal structure prediction was described by Maddox as "one of the stigmas in physics" (Maddox, 1988). In the past decade, machine learning is being used for the prediction of the crystal structure. An abstract idea of heuristic rule extraction was transferred by Curtarolo et al. (2003) to a huge library of information calculated by ab initio method and combined quantum mechanical calculations with machine learning for predicting the binary alloy's crystal structures. Specifically, the model of machine learning was used for predicting various candidate crystal structures, and for determining the thermodynamic stability, DFT calculations were used. Nonetheless, the drawback of this method was that instead of novel structures, only crystal structures present in the database were predicted by it. The structure prediction problem using atomic location points, atomic size, and electronegativity for describing the crystal structure, was investigated by Ceder et al. (2006). Bayesian probability and principal component regression were employed in it for relating atom size and electronegativity to crystal structure. For knowledge extraction point of view, from experimental or computational data, an informatics-based model for structure prediction was constructed by Fischer et al. (2006), DMSP (Data Mining Structure Predictor) that rigidly mines correlations incorporated within the data obtained from the experiment, and efficiently utilizes them to direct the techniques of quantum mechanics toward crystal structures(stable). Two point-clustering algorithms named as DBSCAN and OPTICS were introduced by Phillips and Voth (2013), using these we can automatically identify new crystalline structures from large coordinates datasets. In structure prediction, for addressing high dimensionality challenges of the microstructure space, non-uniqueness of solutions, and multi-objective design requirements, a systematic framework of machine learning was proposed by Liu et al. (2015), it consists classification algorithms, feature selection, and random data generation algorithms, all these predict magneto-elastic Fe-Ga alloy's microstructures. As shown by the results, traditional computational methods were outperformed by this framework, it reduced the average running time by almost 80%. Van Roekeghem et al. (2016) combined regression analysis and PCA with finite-temperature phonon calculations and computed the mechanical stability of cubic perovskite structures with fluorides and semiconducting oxides (approximately 400 fluorides and oxides were calculated) at 1000K, 300K, and 0K. Ultimately, at high temperatures, around 92 mechanically stable perovskite compounds were found, among which 36 were discoveries. For OLEDs, organic light-emitting diodes, focusing on novel emissive layers, a machine learning method was employed by Gómez-Bombarelli et al. (2016) aiming to efficiently screen OLED molecules. In this model, neural networks used for training the algorithm were multitasking and for converting molecules into a fixed-dimensional vector, ECFPs (extended connectivity fingerprints) were used. Through collaborative decision-making and machine learning prescreening 2500 novel OLED molecules were identified from 400,000 molecules. In devices, this excellent predictive power resulted in achieving over 22% efficiency. An LR model was used by Sendek et al. (2017) for lithium-ion conductor (solid) material's screening. The numbers of candidates were shortlisted to 317 from 12,831, which is almost a 92.2% reduction through screening the MP database (the shortlisted candidates were those who satisfied specific requirements) for materials. Then, LR was applied for developing a classification model based on ionic conductivity with which further screening could be done, and finally, they obtained 21 most promising materials, overall observed reduction of 99.8% was observed.

There exists a huge majority of unreported failed or 'dark' chemical reactions from previous studies that are archived in the notebooks of the laboratory that cannot be generally accessed. Although, valuable information is contained in these reactions which determine the boundaries between failure and success, and they can be used in discovering new materials. Full advantage of this data was taken by Raccuglia et al. (2016) to demonstrate an alternative approach that uses reaction data trained SVM-derived

DT algorithm for predicting the reaction results of templated vanadium selenite's crystallization. Traditional human strategies were outperformed by this proposed method, as it successfully predicted the conditions for the organic formation of the new templated inorganic product, having almost an 89% success rate (Sendek et al., 2017).

Table 6.4. Machine Learning Applications Used in the New Material's Discovery

Application description	**ML method**	**Achievement**
Guanidinium ionic liquid's design	ANN	6 new salts (guanidinium)
Finding the missing ternary oxide compounds present in nature	Bayesian	209 new compounds
New compound's discovery with ionic substitutions	Bayesian	20 common ion's substitution rates
Discovering crystals	OPTICS & DBSCAN	The new material finding accelerated
For new material's screening in a composition space that is unconstrained	Bayesian	Almost 4500 new materials
Machine-learning-assisted materials discovery using failed experiments, the discovery of new materials assisted by machine learning	SVM	89% success rate
Material's virtual screening	ANN	failed

6.4.2. Component Prediction

Another way used for discovering new material is component prediction. In short, chemical compositions for forming a compound must be decided by one. In contrast to crystal structure prediction, machine learning is applied more widely in component prediction. The drawbacks for semi-empirical or empirical methods are the limited search space for components and many verification experiments and calculations are required by such searches, this whole thing can affect the discovery progress of new materials very severely. Presently, we can divide the research on component prediction based on machine learning into two basic classes: (1) for a given structure, element's combination recommendation from a group of elements, and (2) for discovering new compounds, ionic substitution's presentation. Whereas without any prior knowledge, only using regression analysis, we can perform crystal structure prediction, while using solving, component prediction proceeds for a posteriori probabilities by using a Bayesian statistical model.

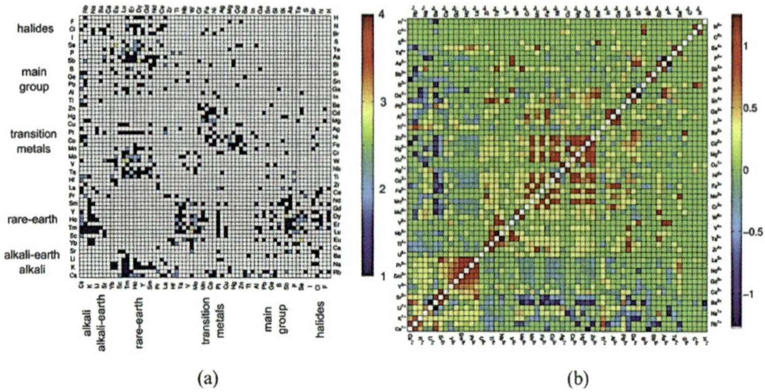

Figure 6.9. (a) New compounds distribution across chemical classes for every A-B-O system, where on the x-axis, A is plotted, and on the y-axis, B is plotted. (b) for each ion couple (a, b) Logarithm (base 10) of the pair correlation gab.

[Source: https://pubs.acs.org/doi/abs/10.1021/ic102031h.]

The difference between Bayesian statistical models and classical statistical models depends on the usage of prior information (Bolstad, 2004). Great importance is paid by the Bayesian statistical model to the usage of single sample information and total available sample information; it also pays attention to the processing, mining, and collection of prior information. These models are used for predicting the material components as for posterior probability estimation; good performance is exhibited by them. A Bayesian statistical method was used by Hautier et al. (2010) for extracting knowledge from around 183 commonly used oxides present in the database of ICSD, as a result, 209 new ternary oxides were successfully predicted. In contrast to the traditionally used exhaustive method, the cost of calculation was reduced by almost a factor of 30. Figure 6.9(a) is showing the distribution of the compounds across chemical classes for each A-B-O system, where A is plotted on the x-axis, and B is plotted on the y-axis. The ordering of the elements is in correspondence to their Mendeleev numbers and this permits us to mark the different chemical classes among which the compounds have been discovered. The same method was used by Meredig et al. (2014) for ternary compound's component prediction, in addition to these oxides, 4500 types of ternary compounds having thermodynamic stability were predicted and in contrast to a single first-principles calculation, the time for computation was also reduced (around 6 orders of magnitude) (Bolstad, 2004).

In a study of component recommendation based on ionic-substitution, a model was proposed by Hautier et al. (2011) for assessing the probability of substitution of ionic species for each other while maintaining the crystal structure; it also showed that how this sort of model can be utilized for predicting the crystal structures and new compounds. The model's predictive power was demonstrated employing cross-validation on quaternary ionic compounds which were 2967 in number and were provided by the ICSD. Figure 6.9(b) shows that a tendency not to substitute is indicated by negative values, while a tendency to substitute is indicated by positive values, and in the model, the embedded substitution rules can be used for predicting the compounds in the lightly dense quaternary space.

6.5. THE MACHINE LEARNING APPLICATIONS USED FOR VARIOUS OTHER PURPOSES

The application of machine learning for new materials discovery and material property prediction has yielded exceptional results. Additionally, it is also practiced in solving materials science-related problems that involve experiments and massive computations. Note that traditional methods cannot solve many of these problems.

6.5.1. Process Optimization

In material synthesis, process optimization serves as the design for process parameters. In the past, rational processing procedures for material were formulated through the accumulation of experience and theoretical analysis. A machine learning method is represented by FNNs, fuzzy neural networks, for integrating fuzzy inference with the neural network's extraordinary learning capability to derive the prime principles for a fuzzy system. FNN and MLR techniques were used by Han et al. (2011) for establishing relationship models between the mechanical properties of the Ti-10V-2Fe-3Al (titanium alloy) and technological parameters; by utilizing these models, we can quickly select the optimal processing parameters required to achieve the desired mechanical properties. In contrast to the MLR method, with the experimental results, a much better agreement is achieved by the FNN method, it shows a relative error less than 7%, similarly, better accuracy of predictions is observed in terms of R^2, MAPE, and RMSE indices. Physically models based on neuro-fuzzy, in Zhu et al. (2003) and Abbod et al. (2002)

were jointly used for predicting the microstructural evolution and flow stress during aluminum alloys thermo-mechanical processing. Moreover, LSSVMs, the least squares support vector machines-based model in Fang et al. (2009) was regarded as an effective and powerful modeling tool to optimize aluminum alloy's aging process.

6.5.2. Finding Density Functionals

Each year, around 10,000 papers report solutions for problems related to electronic structure, caused by using KS DFT (Kohn-Sham). The XC, exchange-correlation energy, is approximated as electronic spin density's functional. The resulting quality is essentially dependent on the density functional approximations, i.e., for strongly correlated systems, the currently used approximations often fail, marking the methodology useless for a few of the interesting problems. By specifying the important technical concepts required for applying machine learning concepts on DFT problems, machine learning was adopted by Snyder et al. (2012) for solving a problem of prototype density functional: spinless fermions (non-interacting ones) confined to a box that is 1dimensional. The preciseness achieved in the system's kinetic energy approximation, having mean absolute errors less than 1 kcal/mol on densities (test ones) concerning the training set, trained on densities less than 100, is far better in proficiency than existing approximations. Furthermore, it is capable of producing accurate self-consistent densities. Many more inputs for achieving chemical accuracy are also used by this MLA (machine learning approximation) approach, but overall this model requires less insight.

6.5.3. Battery Monitoring

During operation, the state of a battery is continuously determined and overall is termed as battery monitoring. As the battery's state is affected by different conditions either external or internal, it is a difficult job in battery management systems (BMSs). Moreover, these conditions and the state of the battery have a nonlinear relationship and it changes over the battery's lifespan. For battery monitoring, the three traditional methods used are Coulomb counting, voltage pulse response, and impedance spectroscopy. Each of these models has similar drawbacks: they are suitable for a specific battery type and are responsible for the estimation of the battery's state of charge (SoC) but only in the steady-state, not when the battery discharges or charges, as it fails there. A superior mean for the prediction of parameters of battery is provided by machine learning, as it has the advantage that it can capture the relationship between the state of the battery and its related factors through the construction of a trained model (Shi et al., 2015). Huge efforts have been made for employing methods of machine learning for monitoring different battery state parameters in real-time, i.e., remaining useful life (RUL), state of health (SoH), impedance parameters, SoC, available power, and capacity.

An FNN model was established by Li et al. (2009) for the estimation of SoC, in which primary input variables used are current, voltage, and internal resistance of the battery. Another machine learning system for SoC estimation was proposed by Lee et al. (2007), this system's implementation was based on cerebellar-model-articulation-controller networks, FNNs, and learning controllers. By using the discharge efficiency and voltage as the input variables, the system generated an estimate showing the available residual power of the battery, and additional useful information was also provided to users. For the battery's current aging state prediction, a BP-ANN optimized model is applied, which uses gradient descent (GD) and the Gauss-Newton method (GN) (Fleischer et al., 2013). The most powerful performance indicator is SoH. It is closely interlinked with the cell/battery's ability for performing a charge function or particular discharge in the discharge-charge-stand-cycle regime, at any given instant. For SoH estimation, SVM methods are a powerful means. Based on virtual standard performance tests and SVM models, the SoH estimation method has been evaluated and validated (Klass et al., 2014). Additionally, for monitoring battery states, extended Kalman filter (EKF) is also combined with machine learning. A neural network was used by Charkhgard and Farrokhi (2010) and the EKF for modeling the SoCs of LIBs, attaining EKF state variables' fast convergence, and achieving better estimation of the SoC.

6.6. COUNTERMEASURES FOR AND ANALYSIS OF COMMON PROBLEMS

6.6.1. Sample Construction

A subset of the data (original), selected to be studied in some prescribed manner is termed as a sample. In the machine learning perspective, basic data is referred by the sample: it mostly includes test data and training data. Currently, we can divide the sample construction problems into three different kinds: sample data source, feature vector's construction, and sample size determination (Addin et al., 2007). In materials science, the sample data, usually originated from experimental measurements and computational simulations, are collected by different schools or research institutions, and in them, a centralized management system usually lacks. This problem is alleviated by materials data infrastructures development, although each database is not unified in the format of data and is separate, still, the machine learning applicability is limited by it (Sundararaghavan & Zabaras, 2005).

Feature vectors are largely responsible for determining the accuracy of the prediction model and are critical. Ideally, a simple physical basis to extract the major chemical and structural trends should be provided by the feature vectors and hence allows fast predictions for chemistries of new material (Fernandez et al., 2014; Martinez et al., 2014). In materials research, the widely used feature vectors mainly include the structure, composition, Coulomb matrix, and electron density. For a specific application, a certain eigenvector is intended; hence there is no unified eigenvector that can be used effectively for all applications. As feature vectors, Coulomb matrices were employed by Hansen et al. (2013) to predict the atomization energies of molecules and demonstrated that the forecasting results are greatly influenced by different representations.

The necessity of materials representation was also emphasized by Schütt et al. (2014) throughout his experiments based problem-solving process. The crucial role of feature vectors was analyzed by Ghiringhelli et al. (2015), employing the Selection Operator (LASSO) and Least Absolute Shrinkage techniques for selecting the features, revealing the systematical process of finding a meaningful descriptor. PCA combined with SVM and DT methods was used by Balachandran et al. (2015) for uncovering the functional forms, intimating the most commonly used features. Moreover, the proposed method was applied by these authors for studying two major classes of materials, termed as, (1) rare earth-main group RM intermetallics and (2) wideband-gap AB compounds, and exhibited that we can use this method for designing new ductile materials.

6.6.2. Generalization Ability

Generalization ability is the ability to accurately predicting the new examples, differing from those which were used for training. For machine learning, this ability is regarded as the crucial evaluation criteria. In SLT, the model bf generalization ability is typically expressed in generalization error terms. Improving the model's generalization ability means to minimize the model's structure risk, and it is the universal aim of machine learning.

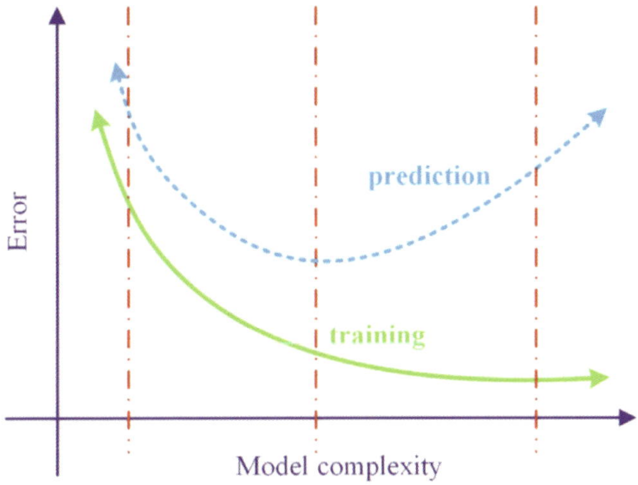

Figure 6.10. Machine learning model's generalization ability.

[Source:https://www.sciencedirect.com/science/article/pii/S2352847817300515.]

As Figure 6.10 shows, as the complexity of the model increases, prediction and training errors tend to differ from each other. The ability of a model to generalize is mostly associated with the problem of over-fitting and/or under-fitting. The definition of under-fitting is that the sample instances (data) were not enough or the learning function was not the right one for the task, alternatively, over-fitting happens when a model becomes too complex, meaning it has too many parameters as compared to the number of observations. The ability of generation gained in various material discovery applications and design differ on a large scale. The root means square error can differ over a huge range, for example, form 0.004467 (model having a good generalization ability) to 20.195 (model having poor generalization ability). Due to this, the procedure to improve the generalization ability of a given model is a crucial problem, and to address this problem quality of the sample, the size of the sample, and the training function that is used should all be considered.

6.6.3. Understandability

Currently, most models of machine learning are considered a "black box," this is because the knowledge extracted from the models is not easy to understand. An example of this is, when using an SVM model for solving regression or classification tasks, the finest classification plane and/ or the fitted curve parameters gained through testing are hidden and invisible in the model. The coherence of knowledge representation is an important measure for assessing a learning algorithm. In many domains, it is expected for a machine learning model to be comprehensible due to its ability to be handled as a model with comprehensive rules and patterns. Models with good comprehensibility are required for machine learning applications in material research. Initially, for predicting the material's behavior, symbolic methods with good comprehensibility were used. Due to the growth of statistical learning methods, a difficulty of poor comprehensibility arises. Thus, the problem that needs to be addressed as soon as possible is the question of how to convert a "black box" into a "white box." To solve such a problem the two common methodologies are: (1) Try developing a more comprehensive algorithm and start avoiding the use of algorithms with poor comprehensibility. Yang et al. introduced a way to study the informative capacity of ANNs, thereby, solving the "black box" problem, in Yang et al. (2005). (2) From the results of a poorly intelligible algorithm, knowledge shall be extracted out. For example, in an intelligible manner, we can express the knowledge present in a neural. Zhou (2004) regarding the extraction of rule from neural networks, presented a review of the on-going rule quality evaluation framework known as the FACC framework, debating that two different rule extraction goals have been confused. A new technique called the impulse-force-based ART (IFART) neural network was introduced by Liu et al. (2004), for predicting earthquake and extracted IF-THEN rules, based on the architecture of trained IFART neural network. Resulting in the improvement of both the reliability and accuracy of earthquake prediction.

6.6.4. Usability

The extent of complications faced when using methods of machine learning for solving a practical problem is called usability. In material sciences, two major aspects regarding the complexity of machine learning are 1) Due to the complexity of machine learning tasks it cannot be done without professional instructions and knowledge, e.g., for increasing the accuracy of material property prediction, one should use the correlation analysis and dimension reduction should be used. A study was reported in Curtarolo et al. (2003) for crystal structure prediction, because of high dimensionality in the sample, PCA was used to reduce the dimensionality of the problem thereby increasing the accuracy of prediction. For explaining the prediction's results of organic polymer material's properties, Pilania et al. (2013) used an analysis based on conditional attribute correlation. 2) It is also a complex task to determine the parameters. Since machine learning algorithms are very reactive to these kernel functions and parameters, it is very important to determine the parameters in the machine learning process. Although mostly through manual changes or through personal experiences the parameters used in the machine learning methods are determined, however, few optimization algorithms are present that can be used for optimizing the parameters. For example, a PSO algorithm was utilized to optimize ANN and SVM parameters in Pei et al. (2013). Thus, it is urgently required to figure out how to enhance the usability of machine learning methods.

6.6.5. Learning Efficiency

The practical application of machine learning is directly related to the performance speed of the machine learning model. Even though an effort is always made to reduce the training and testing time of a model, but it is not possible to achieve both at once. For example, KNN takes less time in training but takes more time in testing whereas neural network models take more time in training and take less time in testing. As of now, this is not a major problem in machine learning applications because the sizes of the samples are quite small almost varying from only a dozen to thousands. Although over time with the development of material genome projects across the world it is expected that material sciences will also step in the era of "big data," at that time, dataset size will increase at a huge rate. A huge dataset size will introduce learning efficiency problems that would require urgent solutions. One proposed solution is to make use of state of the art computing methods, like cloud computing and parallel computing.

REFERENCES

1. Abbod, M. F., Linkens, D. A., Zhu, Q., & Mahfouf, M. (2002). Physically based and neuro-fuzzy hybrid modelling of thermomechanical processing of aluminium alloys. *Materials Science and Engineering: A*, *333*(1–2), 397–408.
2. Addin, O., Sapuan, S. M., Mahdi, E., & Othman, M. (2007). A Naïve-Bayes classifier for damage detection in engineering materials. *Materials & Design*, *28*(8), 2379–2386.
3. Agrawal, A., & Choudhary, A. (2016). Perspective: Materials informatics and big data: Realization of the "fourth paradigm" of science in materials science. *Apl Materials*, *4*(5), 053208.
4. Alder, B. J., & Wainwright, T. E. (1959). Studies in molecular dynamics. I. General method. *The Journal of Chemical Physics*, *31*(2), 459–466.
5. Allen, F. H. (2002). The Cambridge Structural Database: a quarter of a million crystal structures and rising. *Acta Crystallographica Section B: Structural Science*, *58*(3), 380–388.
6. Altun, F., Kişi, Ö., & Aydin, K. (2008). Predicting the compressive strength of steel fiber added lightweight concrete using neural network. *Computational Materials Science*, *42*(2), 259–265.
7. Alzghoul, A., Alhalaweh, A., Mahlin, D., & Bergström, C. A. (2014). Experimental and computational prediction of glass transition temperature of drugs. *Journal of Chemical Information and Modeling*, *54*(12), 3396–3403.
8. Balachandran, P. V., Theiler, J., Rondinelli, J. M., & Lookman, T. (2015). Materials prediction via classification learning. *Scientific Reports*, *5*, 13285.
9. Belsky, A., Hellenbrandt, M., Karen, V. L., & Luksch, P. (2002). New developments in the Inorganic Crystal Structure Database (ICSD): accessibility in support of materials research and design. *Acta Crystallographica Section B: Structural Science*, *58*(3), 364–369.
10. Beran, G. J. (2015). A new era for ab initio molecular crystal lattice energy prediction. *Angewandte Chemie International Edition*, *54*(2), 396–398.
11. Bertinetto, C., Duce, C., Micheli, A., Solaro, R., Starita, A., & Tiné, M. R. (2009). Evaluation of hierarchical structured representations for QSPR studies of small molecules and polymers by recursive neural

networks. *Journal of Molecular Graphics and Modelling*, *27*(7), 797–802.

12. Binder, K., & Girvin, S. M. (1993). The Monte Carlo method in condensed matter physics. *PhT*, *46*(6), 94.
13. Bishop, C. M. (2006). *Pattern recognition and Machine Learning* (Vol.1, pp. 1–22). springer.
14. Blaiszik, B., Chard, K., Pruyne, J., Ananthakrishnan, R., Tuecke, S., & Foster, I. (2016). The Materials Data Facility: Data services to advance materials science research. *JOM*, *68*(8), 2045–2052.
15. Boettinger, W. J., Warren, J. A., Beckermann, C., & Karma, A. (2002). Phase-field simulation of solidification. *Annual Review of Materials Research*, *32*(1), 163–194.
16. Bolstad, W. M. (2004). *Introduction to Bayesian Statistics* (Vol.1, pp. 1–33). John Willey & Sons. Inc., New Jersey.
17. Camacho-Zuniga, C., & Ruiz-Trevino, F. A. (2003). A new group contribution scheme to estimate the glass transition temperature for polymers and diluents. *Industrial & Engineering Chemistry Research*, *42*(7), 1530–1534.
18. Carrera, G. V., Branco, L. C., Aires-de-Sousa, J., & Afonso, C. A. (2008). Exploration of quantitative structure–property relationships (QSPR) for the design of new guanidinium ionic liquids. *Tetrahedron*, *64*(9), 2216–2224.
19. Castin, N., Fernández, J. R., & Pasianot, R. C. (2014). Predicting vacancy migration energies in lattice-free environments using artificial neural networks. *Computational Materials Science*, *84*, 217–225.
20. Cavaliere, P. (2007). Flow curve prediction of an Al-MMC under hot working conditions using neural networks. *Computational Materials Science*, *38*(4), 722–726.
21. Ceder, G., Morgan, D., Fischer, C., Tibbetts, K., & Curtarolo, S. (2006). Data-mining-driven quantum mechanics for the prediction of structure. *MRS Bull*, *31*(12), 981–985.
22. Charkhgard, M., & Farrokhi, M. (2010). State-of-charge estimation for lithium-ion batteries using neural networks and EKF. *IEEE Transactions on Industrial Electronics*, *57*(12), 4178–4187.
23. Chen, B. T., Chang, T. P., Shih, J. Y., & Wang, J. J. (2009). Estimation of exposed temperature for fire-damaged concrete using support vector machine. *Computational Materials Science*, *44*(3), 913–920.

24. Chen, C., Lu, Z., & Ciucci, F. (2017). Data mining of molecular dynamics data reveals Li diffusion characteristics in garnet Li 7 La 3 Zr 2 O 12. *Scientific Reports*, *7*, 40769.
25. Chen, L. Q. (2002). Phase-field models for microstructure evolution. *Annual Review of Materials Research*, *32*(1), 113–140.
26. Chen, L. Q., Chen, L. D., Kalinin, S. V., Klimeck, G., Kumar, S. K., Neugebauer, J., & Terasaki, I. (2015). Design and discovery of materials guided by theory and computation. *npj Computational Materials*, *1*(1), 1–2.
27. Chen, X., Sztandera, L., & Cartwright, H. M. (2008). A neural network approach to prediction of glass transition temperature of polymers. *International Journal of Intelligent Systems*, *23*(1), 22–32.
28. Chonghe, L., Jin, G., Pei, Q., Ruiliang, C., & Nianyi, C. (1996). Some regularities of melting points of AB-type intermetallic compounds. *Journal of Physics and Chemistry of Solids*, *57*(12), 1797–1802.
29. Chonghe, L., Yihao, T., Yingzhi, Z., Chunmei, W., & Ping, W. (2003). Prediction of lattice constant in perovskites of GdFeO3 structure. *Journal of Physics and Chemistry of Solids*, *64*(11), 2147–2156.
30. Cortes, C., & Vapnik, V. (1995). Support-vector networks. *Machine Learning*, *20*(3), 273–297.
31. Curtarolo, S., Morgan, D., Persson, K., Rodgers, J., & Ceder, G. (2003). Predicting crystal structures with data mining of quantum calculations. *Physical Review Letters*, *91*(13), 135503.
32. De'ath, G., & Fabricius, K. E. (2000). Classification and regression trees: a powerful yet simple technique for ecological data analysis. *Ecology*, *81*(11), 3178–3192.
33. Efron, B., & Tibshirani, R. J. (1994). *An Introduction to the Bootstrap* (Vol.1, pp. 1–33). CRC Press.
34. Eminagaoglu, M., & Eren, S. (2010, October). Implementation and comparison of machine learning classifiers for information security risk analysis of a human resources department. In *2010 International Conference on Computer Information Systems and Industrial Management Applications (CISIM)* (Vol.1, pp. 187–192). IEEE.
35. Fang, S. F., Wang, M. P., & Song, M. (2009). An approach for the aging process optimization of Al–Zn–Mg–Cu series alloys. *Materials & Design*, *30*(7), 2460–2467.

36. Fang, S. F., Wang, M. P., Qi, W. H., & Zheng, F. (2008). Hybrid genetic algorithms and support vector regression in forecasting atmospheric corrosion of metallic materials. *Computational Materials Science*, *44*(2), 647–655.

37. Farrusseng, D., Clerc, F., Mirodatos, C., & Rakotomalala, R. (2009). Virtual screening of materials using neuro-genetic approach: Concepts and implementation. *Computational Materials Science*, *45*(1), 52–59.

38. Fernandez, M., Boyd, P. G., Daff, T. D., Aghaji, M. Z., & Woo, T. K. (2014). Rapid and accurate machine learning recognition of high performing metal organic frameworks for CO_2 capture. *The Journal of Physical Chemistry Letters*, *5*(17), 3056–3060.

39. Fischer, C. C., Tibbetts, K. J., Morgan, D., & Ceder, G. (2006). Predicting crystal structure by merging data mining with quantum mechanics. *Nature Materials*, *5*(8), 641–646.

40. Fleischer, C., Waag, W., Bai, Z., & Sauer, D. U. (2013). Adaptive On-line State-of-available-power Prediction of Lithium-ion Batteries. *Journal of Power Electronics*, *13*(4), 516 527.

41. French, R. H., & Bruckman, L. S. (2020). Learnings from developing an applied data science curricula for undergraduate and graduate students. *MRS Advances*, *5*(7), 347–353.

42. Friedman, J. H. (2001). Greedy function approximation: a gradient boosting machine. *Annals of Statistics*, *1*, 1189–1232.

43. Fujimura, K., Seko, A., Koyama, Y., Kuwabara, A., Kishida, I., Shitara, K., ... & Tanaka, I. (2013). Accelerated materials design of lithium superionic conductors based on first-principles calculations and machine learning algorithms. *Advanced Energy Materials*, *3*(8), 980–985.

44. Gajewski, J., & Sadowski, T. (2014). Sensitivity analysis of crack propagation in pavement bituminous layered structures using a hybrid system integrating Artificial Neural Networks and Finite Element Method. *Computational Materials Science*, *82*, 114–117.

45. Gharagheizi, F., Ilani-Kashkouli, P., & Mohammadi, A. H. (2012). A group contribution method for estimation of glass transition temperature ionic liquids. *Chemical Engineering Science*, *81*, 91–105.

46. Ghiringhelli, L. M., Vybiral, J., Levchenko, S. V., Draxl, C., & Scheffler, M. (2015). Big data of materials science: critical role of the descriptor. *Physical Review Letters*, *114*(10), 105503.

47. Gómez-Bombarelli, R., Aguilera-Iparraguirre, J., Hirzel, T. D., Duvenaud, D., Maclaurin, D., Blood-Forsythe, M. A., ... & Markopoulos, G. (2016). Design of efficient molecular organic light-emitting diodes by a high-throughput virtual screening and experimental approach. *Nature materials*, *15*(10), 1120–1127.
48. Gu, G. X., Chen, C. T., & Buehler, M. J. (2018). De novo composite design based on machine learning algorithm. *Extreme Mechanics Letters*, *18*, 19–28.
49. Guo, Z., Malinov, S., & Sha, W. (2005). Modelling beta transus temperature of titanium alloys using artificial neural network. *Computational Materials Science*, *32*(1), 1–12.
50. Guyon, I., & Elisseeff, A. (2003). An introduction to variable and feature selection. *Journal of machine learning research*, *3*(Mar), 1157–1182.
51. Hachmann, J., Olivares-Amaya, R., Atahan-Evrenk, S., Amador-Bedolla, C., Sánchez-Carrera, R. S., Gold-Parker, A., ... & Aspuru-Guzik, A. (2011). The Harvard clean energy project: large-scale computational screening and design of organic photovoltaics on the world community grid. *The Journal of Physical Chemistry Letters*, *2*(17), 2241–2251.
52. Han, Y. F., Zeng, W. D., Shu, Y., Zhou, Y. G., & Yu, H. Q. (2011). Prediction of the mechanical properties of forged Ti–10V–2Fe–3Al titanium alloy using FNN. *Computational Materials Science*, *50*(3), 1009–1015.
53. Hansen, K., Montavon, G., Biegler, F., Fazli, S., Rupp, M., Scheffler, M., ... & Muller, K. R. (2013). Assessment and validation of machine learning methods for predicting molecular atomization energies. *Journal of Chemical Theory and Computation*, *9*(8), 3404–3419.
54. Häse, F., Valleau, S., Pyzer-Knapp, E., & Aspuru-Guzik, A. (2016). Machine learning exciton dynamics. *Chemical Science*, *7*(8), 5139–5147.
55. Hautier, G., Fischer, C. C., Jain, A., Mueller, T., & Ceder, G. (2010). Finding nature's missing ternary oxide compounds using machine learning and density functional theory. *Chemistry of Materials*, *22*(12), 3762–3767.
56. Hautier, G., Fischer, C., Ehrlacher, V., Jain, A., & Ceder, G. (2011). Data mined ionic substitutions for the discovery of new compounds. *Inorganic Chemistry*, *50*(2), 656–663.

57. Hautier, G., Jain, A., & Ong, S. P. (2012). From the computer to the laboratory: materials discovery and design using first-principles calculations. *Journal of Materials Science*, *47*(21), 7317–7340.
58. Helma, C., Cramer, T., Kramer, S., & De Raedt, L. (2004). Data mining and machine learning techniques for the identification of mutagenicity inducing substructures and structure activity relationships of noncongeneric compounds. *Journal of Chemical Information and Computer Sciences*, *44*(4), 1402–1411.
59. Hill, J., Mulholland, G., Persson, K., Seshadri, R., Wolverton, C., & Meredig, B. (2016). Materials science with large-scale data and informatics: unlocking new opportunities. *Mrs Bulletin*, *41*(5), 399–409.
60. Hohenberg, P., & Kohn, W. (1964). Inhomogeneous electron gas physical review 136. *B864*.
61. Hutchinson, J. (2009). Determination of the glass transition temperature: methods correlation and structural heterogeneity. *Journal of Thermal Analysis and Calorimetry*, *98*(3), 579–589.
62. Isayev, O., Oses, C., Toher, C., Gossett, E., Curtarolo, S., & Tropsha, A. (2017). Universal fragment descriptors for predicting properties of inorganic crystals. *Nature Communications*, *8*(1), 1–12.
63. Jain, A., Hautier, G., Ong, S. P., & Persson, K. (2016). New opportunities for materials informatics: resources and data mining techniques for uncovering hidden relationships. *Journal of Materials Research*, *31*(8), 977–994.
64. Jain, A., Ong, S. P., Hautier, G., Chen, W., Richards, W. D., Dacek, S., ... & Persson, K. A. (2013). Commentary: The Materials Project: A materials genome approach to accelerating materials innovation. *Apl Materials*, *1*(1), 011002.
65. Javed, S. G., Khan, A., Majid, A., Mirza, A. M., & Bashir, J. (2007). Lattice constant prediction of orthorhombic ABO3 perovskites using support vector machines. *Computational Materials Science*, *39*(3), 627–634.
66. Joze, H. R. V., & Drew, M. S. (2010, September). Improved machine learning for image category recognition by local color constancy. In *2010 IEEE International Conference on Image Processing* (Vol.1, pp. 3881–3884). IEEE.

67. Kalidindi, S. R., Medford, A. J., & Mcdowell, D. L. (2016). Vision for data and informatics in the future materials innovation ecosystem. *JOM*, *68*(8), 2126–2137.

68. Kirklin, S., Saal, J. E., Meredig, B., Thompson, A., Doak, J. W., Aykol, M., ... & Wolverton, C. (2015). The Open Quantum Materials Database (OQMD): assessing the accuracy of DFT formation energies. *npj Computational Materials*, *1*(1), 1–15.

69. Klass, V., Behm, M., & Lindbergh, G. (2014). A support vector machine-based state-of-health estimation method for lithium-ion batteries under electric vehicle operation. *Journal of Power Sources*, *270*, 262–272.

70. Kohn, W., & Sham, L. J. (1965). Self-consistent equations including exchange and correlation effects. *Physical Review*, *140*(4A), A1133.

71. Kusne, A. G., Gao, T., Mehta, A., Ke, L., Nguyen, M. C., Ho, K. M., ... & Takeuchi, I. (2014). On-the-fly machine-learning for high-throughput experiments: search for rare-earth-free permanent magnets. *Scientific Reports*, *4*(1), 1–7.

72. Larranaga, P., Calvo, B., Santana, R., Bielza, C., Galdiano, J., Inza, I., ... & Robles, V. (2006). Machine learning in bioinformatics. *Briefings in Bioinformatics*, *7*(1), 86–112.

73. Lee, D. T., Shiah, S. J., Lee, C. M., & Wang, Y. C. (2007). State-of-charge estimation for electric scooters by using learning mechanisms. *IEEE Transactions on Vehicular Technology*, *56*(2), 544–556.

74. Li, G., Wang, H., & Yu, Z. (2009, May). New method for estimation modeling of SOC of battery. In *2009 WRI World Congress on Software Engineering* (Vol. 2, pp. 387–390). IEEE.

75. Liu, L., Shi, J., Zhu, F., Du, H., Li, T., Liu, X., & Lu, H. (2017). J Materiomics, 1, 1–33.

76. Liu, R., Kumar, A., Chen, Z., Agrawal, A., Sundararaghavan, V., & Choudhary, A. (2015). A predictive machine learning approach for microstructure optimization and materials design. *Scientific Reports*, *5*(1), 1–12.

77. Liu, R., Yabansu, Y. C., Agrawal, A., Kalidindi, S. R., & Choudhary, A. N. (2015). Machine learning approaches for elastic localization linkages in high-contrast composite materials. *Integrating Materials and Manufacturing Innovation*, *4*(1), 13.

78. Liu, R., Yabansu, Y. C., Yang, Z., Choudhary, A. N., Kalidindi, S. R., & Agrawal, A. (2017). Context aware machine learning approaches

for modeling elastic localization in three-dimensional composite microstructures. *Integrating Materials and Manufacturing Innovation*, *6*(2), 160–171.

79. Liu, Y., Liu, H., Zhang, B., & Wu, G. (2004, August). Extraction of if-then rules from trained neural network and its application to earthquake prediction. In *Proceedings of the Third IEEE International Conference on Cognitive Informatics, 2004.* (Vol.1, pp. 109–115). IEEE.
80. Liu, Y., Wu, J., Yang, G., Zhao, T., & Shi, S. (2019). Predicting the onset temperature (Tg) of GexSe1− x glass transition: a feature selection based two-stage support vector regression method. *Science Bulletin*, *64*(16), 1195–1203.
81. Lufaso, M. W., & Woodward, P. M. (2001). Prediction of the crystal structures of perovskites using the software program SPuDS. *Acta Crystallographica Section B: Structural Science*, *57*(6), 725–738.
82. Maddox, J. (1988). Crystals from first principles. *Nature*, *335*(6187), 201–201.
83. Majid, A., Khan, A., & Choi, T. S. (2011). Predicting lattice constant of complex cubic perovskites using computational intelligence. *Computational Materials Science*, *50*(6), 1879–1888.
84. Majid, A., Khan, A., Javed, G., & Mirza, A. M. (2010). Lattice constant prediction of cubic and monoclinic perovskites using neural networks and support vector regression. *Computational Materials Science*, *50*(2), 363–372.
85. Martinez, R. F., Okariz, A., Ibarretxe, J., Iturrondobeitia, M., & Guraya, T. (2014). Use of decision tree models based on evolutionary algorithms for the morphological classification of reinforcing nano-particle aggregates. *Computational Materials Science*, *92*, 102–113.
86. Meredig, B., Agrawal, A., Kirklin, S., Saal, J. E., Doak, J. W., Thompson, A., ... & Wolverton, C. (2014). Combinatorial screening for new materials in unconstrained composition space with machine learning. *Physical Review B*, *89*(9), 094104.
87. Mitchell, T. M. (1999). Machine learning and data mining. *Communications of the ACM*, *42*(11), 30–36.
88. Mohn, C. E., & Kob, W. (2009). A genetic algorithm for the atomistic design and global optimisation of substitutionally disordered materials. *Computational Materials Science*, *45*(1), 111–117.

89. Mueller, T., Kusne, A. G., & Ramprasad, R. (2016). Machine learning in materials science: Recent progress and emerging applications. *Reviews in Computational Chemistry*, *29*, 186–273.
90. Murphy, K. P. (2012). *Machine Learning: A Probabilistic Perspective* (Vol.1, pp. 1–33). MIT Press.
91. Nantasenamat, C., Isarankura-Na-Ayudhya, C., Naenna, T., & Prachayasittikul, V. (2009). *A Practical Overview of Quantitative Structure-Activity Relationship, 1,* 1–33.
92. National Science and Technology Council (US). (2011). *Materials Genome Initiative for Global Competitiveness* (Vol. 1, pp. 1–33). Executive Office of the President, National Science and Technology Council.
93. Ning, X., Walters, M., & Karypisxy, G. (2012). Improved machine learning models for predicting selective compounds. *Journal of Chemical Information and Modeling*, *52*(1), 38–50.
94. Olson, G. B. (2000). Designing a new material world. *Science*, *288*(5468), 993–998.
95. Olsson, F. (2009). *A Literature Survey of Active Machine Learning in the Context of Natural Language Processing, 1,* 1–32.
96. Paszkowicz, W. (2013). Genetic algorithms, a nature-inspired tool: a survey of applications in materials science and related fields: part II. *Materials and Manufacturing Processes*, *28*(7), 708–725.
97. Peck, R., Olsen, C., & Devore, J. L. (2015). *Introduction to Statistics and Data Analysis* (Vol.1, pp. 1–33). Cengage Learning.
98. Pei, J. F., Cai, C. Z., & Zhu, Y. M. (2013). Modeling and predicting the glass transition temperature of vinyl polymers by using hybrid PSO-SVR method. *Journal of Theoretical and Computational Chemistry*, *12*(03), 1350002.
99. Pei, J. F., Cai, C. Z., Zhu, Y. M., & Yan, B. (2013). Modeling and Predicting the Glass Transition Temperature of Polymethacrylates Based on Quantum Chemical Descriptors by Using Hybrid PSO-SVR. *Macromolecular Theory and Simulations*, *22*(1), 52–60.
100. Phillips, C. L., & Voth, G. A. (2013). Discovering crystals using shape matching and machine learning. *Soft Matter*, *9*(35), 8552–8568.
101. Pilania, G., Wang, C., Jiang, X., Rajasekaran, S., & Ramprasad, R. (2013). Accelerating materials property predictions using machine learning. *Scientific Reports*, *3*(1), 1–6.

102. Provost, F., & Kolluri, V. (1999). A survey of methods for scaling up inductive algorithms. *Data Mining and Knowledge Discovery*, *3*(2), 131–169.

103. Raccuglia, P., Elbert, K. C., Adler, P. D., Falk, C., Wenny, M. B., Mollo, A., ... & Norquist, A. J. (2016). Machine-learning-assisted materials discovery using failed experiments. *Nature*, *533*(7601), 73–76.

104. Rahman, A. (1964). Correlations in the motion of atoms in liquid argon. *Physical Review*, *136*(2A), A405.

105. Raj, R. E., & Daniel, B. S. S. (2008). Prediction of compressive properties of closed-cell aluminum foam using artificial neural network. *Computational Materials Science*, *43*(4), 767–773.

106. Rajan, K. (2005). Materials informatics. *Materials Today*, *8*(10), 38–45.

107. Rao, H. S., & Mukherjee, A. (1996). Artificial neural networks for predicting the macromechanical behaviour of ceramic-matrix composites. *Computational Materials Science*, *5*(4), 307–322.

108. Reich, Y., & Travitzky, N. (1995). Machine learning of material behaviour knowledge from empirical data. *Materials & Design*, *16*(5), 251–259.

109. Rosenblatt, F. (1958). The perceptron: a probabilistic model for information storage and organization in the brain. *Psychological Review*, *65*(6), 386.

110. Russell, S. J., & Stuart, J. (2003). Norvig. *Artificial Intelligence: A Modern Approach*, 1, 111–114.

111. Salahinejad, M., Le, T. C., & Winkler, D. A. (2013). Capturing the crystal: prediction of enthalpy of sublimation, crystal lattice energy, and melting points of organic compounds. *Journal of Chemical Information and Modeling*, *53*(1), 223–229.

112. Saldana, D. A., Starck, L., Mougin, P., Rousseau, B., Ferrando, N., & Creton, B. (2012). Prediction of density and viscosity of biofuel compounds using machine learning methods. *Energy & Fuels*, *26*(4), 2416–2426.

113. Schütt, K. T., Glawe, H., Brockherde, F., Sanna, A., Müller, K. R., & Gross, E. K. (2014). How to represent crystal structures for machine learning: Towards fast prediction of electronic properties. *Physical Review B*, *89*(20), 205118.

114. Scott, D. J., Coveney, P. V., Kilner, J. A., Rossiny, J. C. H., & Alford, N. M. N. (2007). Prediction of the functional properties of ceramic materials from composition using artificial neural networks. *Journal of the European Ceramic Society*, *27*(16), 4425–4435.
115. Sendek, A. D., Yang, Q., Cubuk, E. D., Duerloo, K. A. N., Cui, Y., & Reed, E. J. (2017). Holistic computational structure screening of more than 12000 candidates for solid lithium-ion conductor materials. *Energy & Environmental Science*, *10*(1), 306–320.
116. Shi, S., Gao, J., Liu, Y., Zhao, Y., Wu, Q., Ju, W., ... & Xiao, R. (2015). Multi-scale computation methods: Their applications in lithium-ion battery research and development. *Chinese Physics B*, *25*(1), 018212.
117. Sivasankaran, S., Narayanasamy, R., Ramesh, T., & Prabhakar, M. (2009). Analysis of workability behavior of Al–SiC P/M composites using backpropagation neural network model and statistical technique. *Computational Materials Science*, *47*(1), 46–59.
118. Snyder, J. C., Rupp, M., Hansen, K., Müller, K. R., & Burke, K. (2012). Finding density functionals with machine learning. *Physical Review Letters*, *108*(25), 253002.
119. Stanev, V., Oses, C., Kusne, A. G., Rodriguez, E., Paglione, J., Curtarolo, S., & Takeuchi, I. (2018). Machine learning modeling of superconducting critical temperature. *npj Computational Materials*, *4*(1), 1–14.
120. Steinbach, I. (2009). Phase-field models in materials science. *Modelling and Simulation in Materials Science and Engineering*, *17*(7), 073001.
121. Sumpter, B. G., Vasudevan, R. K., Potok, T., & Kalinin, S. V. (2015). A bridge for accelerating materials by design. *NPJ Computational Materials*, *1*(1), 1–11.
122. Sundararaghavan, V., & Zabaras, N. (2005). Classification and reconstruction of three-dimensional microstructures using support vector machines. *Computational Materials Science*, *32*(2), 223–239.
123. Topcu, I. B., & Sarıdemir, M. (2007). Prediction of properties of waste AAC aggregate concrete using artificial neural network. *Computational Materials Science*, *41*(1), 117–125.
124. Ustun, C., DeFor, T. E., Rashidi, A., Devine, S., Miller, J., & Weisdorf, D. (2017). Time-to-Event Ratio to Predict Outcome in Patients with Acute Myeloid Leukemia Undergoing Allogeneic Hematopoietic Cell Transplantation?. *Biology of Blood and Marrow Transplantation*, *23*(10), 1804–1808.

125. Vahed, A., & Omlin, C. W. (1999, November). Rule extraction from recurrent neural networks using a symbolic machine learning algorithm. In *ICONIP'99. ANZIIS'99 & ANNES'99 & ACNN'99. 6th International Conference on Neural Information Processing. Proceedings (Cat. No. 99EX378)* (Vol. 2, pp. 712–717). IEEE.

126. van Roekeghem, A., Carrete, J., Oses, C., Curtarolo, S., & Mingo, N. (2016). High-throughput computation of thermal conductivity of high-temperature solid phases: the case of oxide and fluoride perovskites. *Physical Review X*, *6*(4), 041061.

127. Vapnik, V. (2013). *The Nature of Statistical Learning Theory (Vol.1, pp. 1–33)*. Springer Science & Business Media.

128. Von Lilienfeld, O. A., Ramakrishnan, R., Rupp, M., & Knoll, A. (2015). Fourier series of atomic radial distribution functions: A molecular fingerprint for machine learning models of quantum chemical properties. *International Journal of Quantum Chemistry*, *115*(16), 1084–1093.

129. Wang, S. C. (2003). Artificial neural network. In *Interdisciplinary Computing in Java Programming* (Vol.1, pp. 81–100). Springer, Boston, MA.

130. Wang, X., Xiao, R., Li, H., & Chen, L. (2017). Quantitative structure-property relationship study of cathode volume changes in lithium ion batteries using ab-initio and partial least squares analysis. *Journal of Materiomics*, *3*(3), 178–183.

131. Ward, L., & Wolverton, C. (2017). Atomistic calculations and materials informatics: A review. *Current Opinion in Solid State and Materials Science*, *21*(3), 167–176.

132. Ward, L., Agrawal, A., Choudhary, A., & Wolverton, C. (2016). A general-purpose machine learning framework for predicting properties of inorganic materials. *npj Computational Materials*, *2*(1), 1–7.

133. Wu, H., Lorenson, A., Anderson, B., Witteman, L., Wu, H., Meredig, B., & Morgan, D. (2017). Robust FCC solute diffusion predictions from ab-initio machine learning methods. *Computational Materials Science*, *134*, 160–165.

134. Wu, S. (2013). A review on coarse warranty data and analysis. *Reliability Engineering & System Safety*, *114*, 1–11.

135. Xu, L., Wencong, L., Chunrong, P., Qiang, S., & Jin, G. (2009). Two semi-empirical approaches for the prediction of oxide ionic

conductivities in ABO3 perovskites. *Computational materials Science*, *46*(4), 860–868.

136. Yang, L., Wang, P., Jiang, Y., & Chen, J. (2005). Studying the Explanatory Capacity of Artificial Neural Networks for Understanding Environmental Chemical Quantitative Structure–Activity Relationship Models. *Journal of Chemical Information and Modeling*, *45*(6), 1804–1811.
137. Yu, K., & Cheng, Y. (2007). Machine learning techniques for the prediction of the peptide mobility in capillary zone electrophoresis. *Talanta*, *71*(2), 676–682.
138. Yu, X. L., Yi, B., & Wang, X. Y. (2008). Prediction of the glass transition temperatures for polymers with artificial neural network. *Journal of Theoretical and Computational Chemistry*, *7*(05), 953–963.
139. Zhang, X. J., Chen, K. Z., & Feng, X. A. (2008). Material selection using an improved genetic algorithm for material design of components made of a multiphase material. *Materials & Design*, *29*(5), 972–981.
140. Zhou, Z. H. (2004). Rule extraction: Using neural networks or for neural networks? *Journal of Computer Science and Technology*, *19*(2), 249–253.
141. Zhu, Q., Abbod, M. F., Talamantes-Silva, J., Sellars, C. M., Linkens, D. A., & Beynon, J. H. (2003). Hybrid modelling of aluminium–magnesium alloys during thermomechanical processing in terms of physically-based, neuro-fuzzy and finite element models. *Acta Materialia*, *51*(17), 5051–5062.

Chapter 7

Machine Learning and Quantum Physics

CONTENTS

7.1. INTRODUCTION

We already stablished machine learning as a discipline that seeks for building a framework able to formulate predictions about data and find worthwhile information. Is the root of artificial intelligence (AI) and has induced various aspects of the latest technologies, from natural language processing and face recognition to automated self-driving cars (Hush, 2017; Sarma et al., 2019).

The discipline is expeditiously growing, and pervasively applicable. Machine learning is used by Google Translate's online service that converts Chinese characters into English text without any human interference. A robot that has beaten the world's finest players in Go (an archaic board game) is AlphaGo that was built using machine-learning techniques. It becomes proficient in the game and is considered to be the highest AI accomplishment (Mitchell & Thrun, 1993; LeCun et al., 2015). The game was universally believed to be too complicated for machines to come through due to its large number of possible actions until AlphaGo exposed its excellence (Gao et al., 2018; Schuld, 2018).

The curse of dimensionality is one of the biggest challenges confronted by machine learning. Generally, the machine requires a vast range of training data sets exponential in the dimension d to learn wanted information. If a data set belongs to high-dimensional space, then it promptly evolves into computationally chaotic. Quantum mechanics is identical to that complexity, for which normally an exponential quantity of information is needed as well to completely define a quantum many-body state (Li et al., 2015; Carleo & Troyer, 2017).

The most favorable quantitative theory of nature perhaps is the quantum theory. It determines the ground for understanding partially physics on full-length scales, from elementary particles, e.g., quarks and electrons to some important aspects of gigantic objects, e.g., galaxies and stars formation, and also make a base for modern technologies covering from transistors and lasers to quantum computers and nuclear magnetic resonators (Nielsen & Chuang, 2002; Silver et al., 2017). It can be asked from the great accomplishments of both quantum physics and machine learning: Can these both seemingly independent but closely connected disciplines incorporate in a seamless, corporative way?

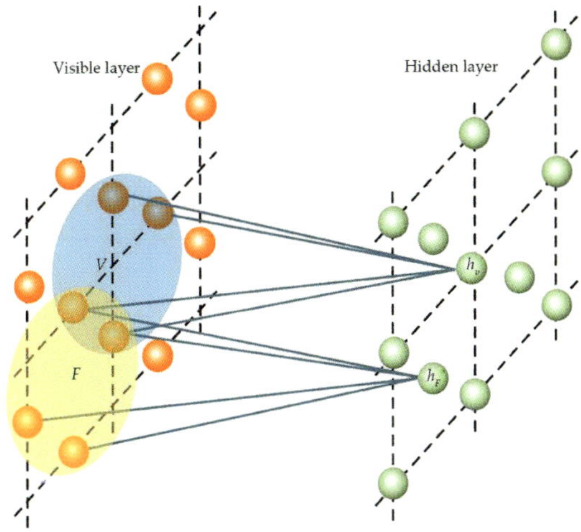

Figure 7.1. The Restricted-Boltzmann-Machine portrayal of the toric code state using intrinsic topological order. There are four visible neurons for each face for vertex v that are associated with one hidden neuron hf or hv. The portrayal is effective due to each connection associated with one parameter in the neural network, thus the number of parameters grows linearly with the size of the system rather than exponentially.

[Source: https://physicstoday.scitation.org/doi/pdf/10.1063/PT.3.4164.]

It seems close to science fiction, but fusion technologies are occurring currently and may result in mind-boggling progress in both disciplines. Over the previous two decades, machine learning has proceeded productively, and many challenges that were extensive or even unreachable to automated learning have been resolved now. New potentials for machine learning are promoted by that progress to solve open challenges in quantum physics (Rupp, 2015; Gupta et al., 2017). At the same time, implementations of computation in physical theories have been revolutionized by the concept of quantum information processing. Tantalizing prospects might be proposed by new quantum algorithms to improve machine learning itself. The cooperation between quantum physics and machine learning will certainly give an advantage to both disciplines (Schuld et al., 2015).

7.2. UNCOVERING PHASES OF MATTER

Supervised learning is used in the application of machine learning to problems of physics, where an algorithm is instructed with beforehand labeled data;

the target of that algorithm is to capture that information and build up a general rule for labeling data that are beyond the training set (Carrasquilla & Melko, 2017). As an example, to identify images of cats and dogs, thousands of pictures of cats and dogs will be taken by supervised learning algorithms along with their labels and establishing a relationship between their labels and pixel values of images. Then those labels are assigned to unseen images (Broecker et al., 2017; Venderley et al., 2018).

The same supervised learning strategy can be used to recognize unique stages of matter and their transitions, one of the key subjects in condensed-matter physics. This idea was first examined by Juan Carrasquilla and Roger Melko (2017) in their research of the ferromagnetic Ising model, which characterizes discrete atomic spins organized on a lattice. The spins show an ordered ferromagnetic state at low temperatures and a disordered paramagnetic state at high temperatures, and at some critical temperature $T_{c,}$ a state transition takes place between the two (Wetzel & Scherzer, 2017).

Rather than sorting cats and dogs, Carrasquilla and Melko (2017) trained an algorithm using equilibrium spin configurations collected from Monte Carlo simulations to identify ferromagnetic and paramagnetic phases. It has been exhibited that by training algorithm with those labeled datasets, it could accurately label the unseen dataset. Furthermore, T_c was discovered by looking over a range of temperatures, and the critical exponents were found that are vital to the study of state transitions (Deng et al., 2017; Ponte & Melko, 2017).

Supervised learning demands that users should know earlier the labels of their data. On the other hand, unlabeled training samples are used by unsupervised learning which enables the network to find structures and meaningful patterns in them. Clustering is one of the popular techniques of unsupervised learning that divides training samples into various groups based on similarities between them and those groups are used to label new unseen data (Van Nieuwenburg et al., 2017; Zdeborová, 2017). Clustering was applied to the Ising model by Lei Wang (2016) in which ferromagnetic and paramagnetic states and the transition between them are identified successfully, regardless of giving explicit sorting criteria to the algorithm.5 Meanwhile, a confusion scheme was proposed by Evert van Nieuwenburg and his coworkers that merged supervised and unsupervised learning both (Wang, 2016). Their approach was tested on various models including the Ising model which expressed that it could detect several states and the transitions between them (Broecker et al., 2017; Rodriguez-Nieva & Scheurer, 2019).

7.3. NEURAL-NETWORK REPRESENTATION

With the rapid progress of machine-learning algorithms for detecting the state of matter, an artificial neural network has been made an exciting development to depict quantum states and solve corresponding quantum many-body problems.

An exponential amount of information is required by fully describing an arbitrary many-body state in quantum mechanics. Assume a system having N qubits, or quantum bits. Every qubit has two available independent configurations, i.e., 0 or 1; hence $2N$ possible configurations collectively. Computationally, $2N$ complex numbers are required by fully describing the related quantum state (Yang et al., 2015; Pan et al., 2016).

The exponential complexity offers a tremendous threat for numerical simulations of quantum many-body systems achieved on a classical computer, i.e., massively large memory is required for representing even a few qubits. As an illustration, tens of gigabytes are required to simulate a quantum system with 30 qubits that is the largest memory for a personal desktop; tens of petabytes are required to simulate 50 qubits that are more than the memory of the current largest supercomputer in the world; and higher bytes than the number of atoms in the apparent universe are required to simulate 300 qubits (Weiss et al., 2015).

Luckily, only a limited edge of the whole Hilbert space of quantum states is typically accessed by most physical states of interest, e.g., the ground states of many-body Hamiltonians, and for that reason can be elaborated with a shorter amount of information. Therefore, to deal with quantum many-body problems using classical computers, it is essential to design pointed representations of those phases such that they hold their crucial physical properties.

Tensor-network representation is the famous description for each phase, in which each qubit assigns a tensor, and jointly those tensors elaborate the many-body quantum state. The most physical states can be represented by such a construction efficiently such that the information quantity needed with the size of system scales only polynomially instead of exponentially (Van Nieuwenburg et al., 2017).

Compact representations of quantum states can be constructed using artificial neural networks which is an exceptionally abstracted and simplified design of the human brain. The idea was first proposed by Giuseppe Carleo and Matthias Troyer (2017) while introducing a new representation formed on the restricted Boltzmann machine (RBM) which is a remarkable neural network extensively used in the machine-learning association (Schollwöck, 2011). As shown in figure 8.1, an RBM is organized as two layers of neurons, i.e., (i) a visible, and (ii) a hidden layer. The visible neurons indicate the physical qubits, whereas the hidden neurons depict the auxiliary degrees of freedom that are reduced by a summation eventually to generate the output of the network, a complex number that attends as the coefficient for the matching qubit configuration (Tu et al., 2016).

A question arises, what types of quantum many-body phases can be described efficiently by RBMs? RBMs represent well certain exotic phases like topological states (Deng et al., 2017). Figure 8.1 illustrates the RBM description for the ground state of the toric-code Hamiltonian, a topological state which was proposed by Alexei Kitaev with the framework of topological quantum computation. Each hidden neuron of RBM associates only to its neighboring four visible neurons to illustrate the toric-code state. Single network parameter describes each association; thus, the total number of parameters is approximately four times the number of qubits with the size of the system which measures linearly instead of exponentially. The conspicuously compact description of the toric-code state can proceed to the excited states also (Deng et al., 2017a; 2017b; Gao & Duan, 2017).

The quantum states with physical interest also exist that proceed with no efficient RBM representation. Nevertheless, if an extra hidden layer is included, then it increases the RBM's relevance. Around all physical quantum states can be represented efficiently with the demanded number of parameters measures merely polynomially with the size of the system by the resultant neural network that is called deep Boltzmann machine.

7.4. ENTANGLEMENT IN NEURAL-NETWORK STZATES

What, then, restricts neural networks in describing quantum many-body states efficiently? The key to the traditional tensor-network representation is quantum entanglement. Is it also a demanding aspect of the neural-network representation?

A physical phenomenon which would influenced the state of a particle instantaneously by the measurements on another particle, indeed when a large distance separates the particles spatially is called Quantum entanglement—Albert Einstein named this phenomenon as "spooky action at a distance."

Entanglement is at the focus of the renowned Schrödinger's cat paradox as well. Quantum entanglement deeply troubled Erwin Schrödinger and Einstein both (Chen et al., 2018; Torlai et al., 2018).

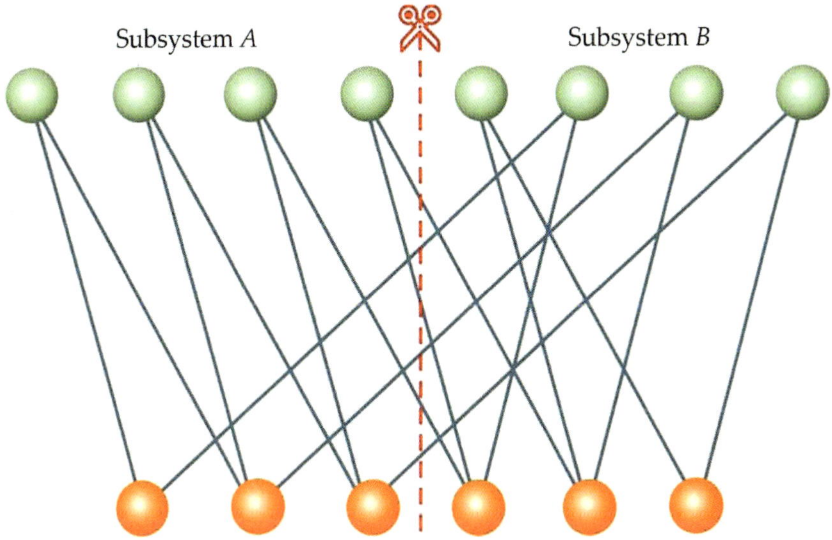

Figure 7.2. A representation of neural-network of a one-dimensional quantum state having maximum volume-law complication: If the system is split into two subsystems, i.e., A and B, the entropy of every subsystem is proportional to its size. At most three hidden neurons are connected to each visible neuron hence, the number of parameters required to define the subsystem grows linearly instead of exponentially with the system volume, as in the representation of a conventional tensor-network.

[Source: https://physicstoday.scitation.org/doi/pdf/10.1063/PT.3.4164.]

As illustrated in Figure 8.2, assuming to decompose a pure quantum many-body state into two subsystems, i.e., *A* and *B*. As entropy can characterize a traditional many-body system, similarly, entanglement entropy can characterize a quantum many-body system. The entanglement area law is satisfied by various natural quantum systems, which means that the entanglement entropy of a subsystem adjusts as only at the surface area or the horizon of the subsystem instead of as its volume (Jia et al., 2019; Choo et al., 2020). For example, Bekenstein– Hawking entropy of a black hole measures as the area of its event border. The source of the black hole entropy is broadly supposed to be the quantum entanglement between the outside and inside of the black hole. The entanglement area law is also satisfied by the ground states of numerous classical local Hamiltonians in quantum many-body physics, however, a rigid proof of that is notably challenging and remains anonymous. In the tensor-network representation of quantum many-body states, entanglement area law is imperative and builds the backbone of great tensor-network-based algorithms. Generally, a tensor network needs a polynomial number of parameters with the size of the system to represent a quantum state that assures the entanglement area law. Hence an efficient tensor-network representation is generally endured by such quantum states. Nevertheless, the conventional tensor-network representation is not efficient for quantum states with enormous entanglement like extremely excited states of quantum Hamiltonians having volume-law entanglement—the exponential number of parameters are required with the size of the system (Bausch & Leditzky, 2020).

The entanglement area law is obeyed by all the RBM neural-network states accompanying short-range connectivity, free of subsystem geometric details, and dimensionality (Gao & Duan, 2017). The area law must be then obeyed by the toric-code states, in which every neuron associates only to its four neighboring vertices, a conviction that has also been approved through other intricated mathematical approaches. An entanglement volume law is satisfied by general RBM states without the short-range provision. The families of RBM states can be constructed analytically with the maximum entanglement. A model of such a construction is portrayed in figure 8.2, from which a compelling inference follows at once: The RBM representation of heavily entangled states is exceptionally efficient. At most three hidden neurons are connected to every visible neuron, therefore linear number of parameters scales with the size of the system; that scaling manifests the superlative power of neural networks in representing quantum many-body states having large entanglement. In contrast, the conventional tensor-network representation requires an exponentially large number of parameters to portray highly entangled states. Undoubtedly, entanglement is not the confined aspect for the efficiency of the neural-network representation (Evangelista, 2014).

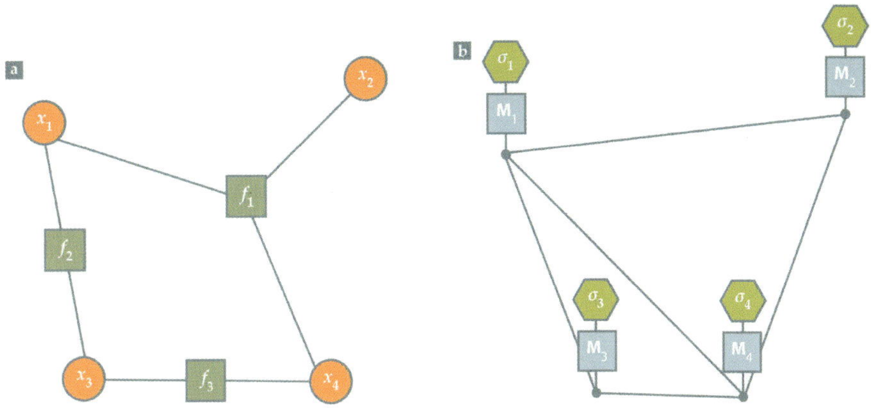

Figure 7.3. Classical and Quantum Generative Models are broadly utilized in both unsupervised and supervised machine learning.

(a) This demonstration of a classical generative model or factor graph illustrates the consolidated probability distribution of observables xi being a product of factor functions: P(x1, x2, x3, x4) = f1(x1, x2, x4) f2(x1, x3) f3(x3, x4). After that, a classical generative learning task is diminished to optimizing the flexible parameters in the factor functions fi.
*(b) It demonstrates a quantum generative model having four qubits σi. The figure indicates a special quantum state established by assisting the two-by-two invertible matrices **Mi** on a tensor network state. Then the probability distribution can be acquired from projective measurements on the emerging state. After that, a quantum generative learning task is diminished to optimizing the flexible parameters in the matrices **Mi**.*

[Source: https://physicstoday.scitation.org/doi/pdf/10.1063/PT.3.4164.]

7.5. QUANTUM MANY-BODY PROBLEMS

Usually, solving quantum many-body problems results in finding either the dynamics of the system's time evolution of the system's ground state. That can be gained via an RBM-based variational learning algorithm selected by Carleo and Troyer in the same study in which the RBM representation has been introduced. The approach was tested on two prototypical quantum spin models—(i) the antiferromagnetic Heisenberg model and (ii) the Ising model in a transverse magnetic field—and discovered that the RBM approach conscientiously grabbed the time evolution for each one and the ground state (White, 2002; Gezzi et al., 2007).

Heavily entangled states are represented by the unusual ability of neural networks that provides a new approach to solve complicated many-body problems involving large entanglement; such problems are unsolvable with traditional methods even or challenging (Cherednik, 1994; Hikami, 1996). The system's ground state is found by the RBM-based variational learning algorithm while applying to a model Hamiltonian with long-range interactions, that has been scientifically proved to carry power-law entanglement. Furthermore, quantum state tomography has also used the RBM technique for highly entangled states, where quantum state tomography is a method of reorganizing the quantum state from the outcomes of quantum measurements.

Another enigmatic characteristic of quantum mechanics is nonlocality which is closely associated with entanglement. Any local realistic characterization of our earth is inhibited by quantum nonlocality as resoundingly established by John Bell. It also indicates the extremely intense departure of the quantum world from the traditional (White, 2002; Cohen et al., 2005). Practically, nonlocality is an imperative resource for numerous devices, i.e., independent quantum technologies, e.g., certifiable random-number generation and secure cryptographic key distribution. For a generic many-body system, the whole description of quantum nonlocality is overly challenging; however, an auspicious approach for at least partly solving that problem is machine learning, specifically RBM-based variational learning.

7.6. QUANTUM-ENHANCED MACHINE LEARNING

The above-mentioned examples have straightforwardly exposed the unparalleled power of machine learning methods to solve numerous challenging quantum problems. Markedly, in contrast to conventional tensor-network strategy, entanglement is not the restricting factor for the related algorithms to learn like representation and for the efficiency of the neural-network representation. Additionally, high-dimensional systems are dealt with neural-network approaches due to the tremendous flexibility of neural-network structures. The opposite holds as well: Quantum technologies, particularly quantum computing, can give a great lift to machine learning. Machine learning generally handles a large amount of data. Fast Fourier transform (FFT) is one of the popular data- analysis techniques. A quantum version of FFT with quantum computers are exponentially speedy than the conventional version. On the other hand, machine-learning algorithms generally demand to solve a large number of linear problems that add up to do many matrix multiplications. Since quantum mechanics is typically described by linear algebra, in carrying out those operations' quantum computers have intrinsic advantages. Max Born, Pascual Jordan, and Werner Heisenberg formulate quantum mechanics which is earlier known as matrix mechanics. In this manner, there are various conceptual connections between quantum computing and machine learning (Dunjko et al., 2016; Havlíček et al., 2019).

Quantum computers aren't anticipated to boost up every machine learning algorithm. Nevertheless, several quantum algorithms have been found by scientists that guarantee exponential speed enhanced for special important tasks (Deng, 2018). HHL algorithm is an algorithm that is elemental to the present quantum machine learning mini-revolution named after its creators Aram **H**arrow, Avinatan **H**assidim, and Seth **L**loyd. Other quantum learning algorithms either use it as a subroutine or extend HHL. A system of linear equations is solved using this algorithm (Hentschel & Sanders, 2010; Escher et al., 2011):

The target of HHL is to solve $\mathbf{Ax} = \mathbf{b}$ for $\mathbf{x}$ given a vector $\mathbf{b}$ and an $N \times N$ matrix $\mathbf{A}$. To output a quantum state, the HHL algorithm reach on the order of log2 N quantum steps for many matrices of physical interest, while the well-known traditional algorithm needs on the order of N log N steps. Its potential benefits might be nullified by various warnings to the HHL algorithm and its variants. For example, the algorithm demands quantum qRAM, or RAM to map a classical vector to a quantum state that could be exponentially costly. A quantum generative model relaxed the qRAM requirement that is recently proposed by Gao & Duan (2017) and hence in the initial step of passing on classical data to quantum states, the problem of exponential overhead is prevented (Biamonte et al., 2017; Dunjko & Briegel, 2018). A different approach is taken by generative models to solve problems via a machine learning as compared to more familiar discriminative models. Regard the previous examples of pictures of cats and dogs to understand the difference between the two. The goal of a discriminative model is to learn the features that figure out the differences between the pictures of animals. The objective of generative models is to be able to generate new figures of cats and dogs. Practically, a latent probability distribution is figured out by the generative approach from a set of training data. In the classical scheme, a factor graph can represent the probability distribution. Still, a quantum state describes the probability distribution for the quantum generative model. Figure 8.3 shows the portrayal of both the quantum and classical generative models. Three significant views show that the quantum generative model has exponential benefits over its classical counterpart. The quantum algorithm can represent more probability distributions efficiently, as well as it is exponentially speedy than the classical one both at generating new data and at learning certain probability distributions. New gates are opened for the quantum generative model to seek the potential of quantum computing to solve challenging machine learning problems, and therefore it should have valuable applications forthcoming (Uvarov et al., 2019).

Machine learning, particularly AI tasks might be significantly breakthrough the above-given examples are just a flash into an increasing sea of quantum algorithms. Other compelling algorithms like quantum support-vector machine and quantum principal component analysis, offer great speed-up potentials. Furthermore, the beginning of offering intriguing merits is a currently presented quantum-inspired tensor-network algorithm for machine learning (Ciliberto et al., 2018).

7.7. FUTURE PARTNERSHIP

Exciting progress has been made in the interdisciplinary field of joining quantum physics and machine learning and expanding swiftly. The above discussions are like a water drop in an ocean (Mehta & Schwab, 2014; DeBenedictis, 2018).

Answers to two vital questions are required while applying machine learning to quantum physics: (1) Can machine learning aid to reveal new physics in quantum systems? and (2) What is the key application for machine learning to solve quantum problems? A learning algorithm is an ambitious project that could answer these two questions promptly because it specialized in determining high-T_c superconductors. New high-T_c superconducting materials could be able to predict and provide new intuitiveness to the theory of superconductivity after training on the abundant available experimental data (Stoudenmire & Schwab, 2016; Carleo & Troyer, 2017).

A unified quantum learning theory has not been progressed for quantum-enhanced machine learning, and some essential questions arise:

- How can large quantum datasets that may ultimately be available, be analyzed by a quantum computer efficiently?

- What learning problems can be solved efficiently by a quantum computer but not by a classical one?
- What is the general criterion for determining if a quantum computer can expedite a machine learning task significantly?

There is an exact map between deep learning and the variational renormalization group method in physics that is an iterative coarse-graining strategy that extracts suitable features for a physical system at various length scales in traditional machine learning, and that map gives a worthwhile vision that why deep learning is powerful (Harrow et al., 2009). Is it feasible to build up such a map in the case of quantum deep learning? Furthermore, a significant milestone would be a boost of practical machine learning tasks by a smoking-gun experimental demonstration of quantum (Steane & Rieffel, 2000; Melko et al., 2019).

It is troublesome yet to predict the future of the quantum and difficult to forecast that when the first practical quantum computer will be introduced. Still, one thing is for sure: The merger of quantum physics and machine learning is a harmonious relationship that could mold them both (Albertsson et al., 2018; Gao et al., 2018).

REFERENCES

1. Albertsson, K., Altoe, P., Anderson, D., Anderson, J., Andrews, M., Espinosa, J. P. A., ... & Bonacorsi, D. (2018). Machine learning in high energy physics community white paper, 1, 1–30. *arXiv preprint arXiv:1807.02876.*
2. Bausch, J., & Leditzky, F. (2020). Quantum codes from neural networks. *New Journal of Physics*, *22*(2), 023005.
3. Biamonte, J., Wittek, P., Pancotti, N., Rebentrost, P., Wiebe, N., & Lloyd, S. (2017). Quantum machine learning. *Nature*, *549*(7671), 195–202.
4. Broecker, P., Assaad, F. F., & Trebst, S. (2017). Quantum phase recognition via unsupervised machine learning, 1, 1–20. *arXiv preprint arXiv:1707.00663.*
5. Broecker, P., Carrasquilla, J., Melko, R. G., & Trebst, S. (2017). Machine learning quantum phases of matter beyond the fermion sign problem. *Scientific reports*, *7*(1), 1–10.
6. Carleo, G., & Troyer, M. (2017). Solving the quantum many-body problem with artificial neural networks. *Science*, *355*(6325), 602–606.
7. Carrasquilla, J., & Melko, R. G. (2017). Machine learning phases of matter. *Nature Physics*, *13*(5), 431–434.
8. Chen, J., Cheng, S., Xie, H., Wang, L., & Xiang, T. (2018). Equivalence of restricted Boltzmann machines and tensor network states. *Physical Review B*, *97*(8), 085104.
9. Cherednik, I. (1994). Integration of quantum many-body problems by affine Knizhnik-Zamolodchikov equations. *Advances in Mathematics*, *106*(1), 65–95.
10. Choo, K., Mezzacapo, A., & Carleo, G. (2020). Fermionic neural-network states for ab-initio electronic structure. *Nature communications*, *11*(1), 1–7.
11. Ciliberto, C., Herbster, M., Ialongo, A. D., Pontil, M., Rocchetto, A., Severini, S., & Wossnig, L. (2018). Quantum machine learning: a classical perspective. *Proceedings of the Royal Society A: Mathematical, Physical and Engineering Sciences*, *474*(2209), 20170551.
12. Cohen, G., Gull, E., Reichman, D. R., & Millis, A. J. (2015). Taming the dynamical sign problem in real-time evolution of quantum many-body problems. *Physical review letters*, *115*(26), 266802.

13. DeBenedictis, E. P. (2018). A future with quantum machine learning. *Computer*, *51*(2), 68–71.
14. Deng, D. L. (2018). Machine learning detection of bell nonlocality in quantum many-body systems. *Physical review letters*, *120*(24), 240402.
15. Deng, D. L., Li, X., & Sarma, S. D. (2017a). Machine learning topological states. *Physical Review B*, *96*(19), 195145.
16. Deng, D. L., Li, X., & Sarma, S. D. (2017b). Quantum entanglement in neural network states. *Physical Review X*, *7*(2), 021021.
17. Dunjko, V., & Briegel, H. J. (2018). Machine learning & artificial intelligence in the quantum domain: a review of recent progress. *Reports on Progress in Physics*, *81*(7), 074001.
18. Dunjko, V., Taylor, J. M., & Briegel, H. J. (2016). Quantum-enhanced machine learning. *Physical review letters*, *117*(13), 130501.
19. Escher, B. M., de Matos Filho, R. L., & Davidovich, L. (2011). General framework for estimating the ultimate precision limit in noisy quantum-enhanced metrology. *Nature Physics*, *7*(5), 406–411.
20. Evangelista, F. A. (2014). A driven similarity renormalization group approach to quantum many-body problems. *The Journal of chemical physics*, *141*(5), 054109.
21. Guo, J., Qiao, L. F., Jiao, Z. Q., Ma, Y. C., Hu, C. Q., Ren, R. J., ... & Jin, X. M. (2018). Experimental machine learning of quantum states. *Physical review letters*, *120*(24), 240501.
22. Gao, X., & Duan, L. M. (2017). Efficient representation of quantum many-body states with deep neural networks. *Nature communications*, *8*(1), 1–6.
23. Gao, X., Zhang, Z. Y., & Duan, L. M. (2018). A quantum machine learning algorithm based on generative models. *Science advances*, *4*(12), eaat9004.
24. Gezzi, R., Pruschke, T., & Meden, V. (2007). Functional renormalization group for nonequilibrium quantum many-body problems. *Physical Review B*, *75*(4), 045324.
25. Gupta, S., Mohanta, S., Chakraborty, M., & Ghosh, S. (2017, August). Quantum machine learning-using quantum computation in artificial intelligence and deep neural networks: Quantum computation and machine learning in artificial intelligence. In *2017 8th Annual*

Industrial Automation and Electromechanical Engineering Conference (IEMECON) (Vol.1, pp. 268–274). IEEE.

26. Harrow, A. W., Hassidim, A., & Lloyd, S. (2009). Quantum algorithm for linear systems of equations. *Physical review letters, 103*(15), 150502.
27. Havlíček, V., Córcoles, A. D., Temme, K., Harrow, A. W., Kandala, A., Chow, J. M., & Gambetta, J. M. (2019). Supervised learning with quantum-enhanced feature spaces. *Nature, 567*(7747), 209–212.
28. Hentschel, A., & Sanders, B. C. (2010). Machine learning for precise quantum measurement. *Physical review letters, 104*(6), 063603.
29. Hikami, K. (1996). Dunkl operator formalism for quantum many-body problems associated with classical root systems. *Journal of the Physical Society of Japan, 65*(2), 394–401.
30. Hush, M. R. (2017). Machine learning for quantum physics. *Science, 355*(6325), 580–580.
31. Jia, Z. A., Yi, B., Zhai, R., Wu, Y. C., Guo, G. C., & Guo, G. P. (2019). Quantum neural network states: A brief review of methods and applications. *Advanced Quantum Technologies, 2*(7–8), 1800077.
32. LeCun, Y., Bengio, Y., & Hinton, G. (2015). Deep learning. *Nature, 521*(7553), 436–444.
33. Li, Z., Kermode, J. R., & De Vita, A. (2015). Molecular dynamics with on-the-fly machine learning of quantum-mechanical forces. *Physical Review Letters, 114*(9), 096405.
34. Mehta, P., & Schwab, D. J. (2014). An exact mapping between the variational renormalization group and deep learning, 1, 1–20. *arXiv preprint arXiv:1410.3831.*
35. Melko, R. G., Carleo, G., Carrasquilla, J., & Cirac, J. I. (2019). Restricted Boltzmann machines in quantum physics. *Nature Physics, 15*(9), 887–892.
36. Mitchell, T. M., & Thrun, S. B. (1993). Explanation-based neural network learning for robot control. In *Advances in Neural Information Processing Systems* (Vol. 1, pp. 287–294).
37. Nielsen, M. A., & Chuang, I. (2002). Quantum computation and quantum information, 1, 1–33.
38. Pan, S., Wu, J., Zhu, X., Zhang, C., & Wang, Y. (2016). Tri-party deep network representation. *Network, 11*(9), 12.

39. Ponte, P., & Melko, R. G. (2017). Kernel methods for interpretable machine learning of order parameters. *Physical Review B, 96*(20), 205146.
40. Rodriguez-Nieva, J. F., & Scheurer, M. S. (2019). Identifying topological order through unsupervised machine learning. *Nature Physics, 15*(8), 790–795.
41. Rupp, M. (2015). Machine learning for quantum mechanics in a nutshell. *International Journal of Quantum Chemistry, 115*(16), 1058–1073.
42. Sarma, S. D., Deng, D. L., & Duan, L. M. (2019). Machine learning meets quantum physics, 1, 1–22. *arXiv preprint arXiv:1903.03516.*
43. Schollwöck, U. (2011). The density-matrix renormalization group in the age of matrix product states. *Annals of physics, 326*(1), 96–192.
44. Schuld, M. (2018). *Supervised learning with quantum computers (Vol.1, pp. 1–33).* Springer.
45. Schuld, M., Sinayskiy, I., & Petruccione, F. (2015). An introduction to quantum machine learning. *Contemporary Physics, 56*(2), 172–185.
46. Silver, D., Schrittwieser, J., Simonyan, K., Antonoglou, I., Huang, A., Guez, A., ... & Chen, Y. (2017). Mastering the game of go without human knowledge. *nature, 550*(7676), 354–359.
47. Steane, A. M., & Rieffel, E. G. (2000). Beyond bits: The future of quantum information processing. *Computer, 33*(1), 38–45.
48. Stoudenmire, E., & Schwab, D. J. (2016). Supervised learning with tensor networks. In *Advances in Neural Information Processing Systems* (Vol. 1, pp. 4799–4807).
49. Torlai, G., Mazzola, G., Carrasquilla, J., Troyer, M., Melko, R., & Carleo, G. (2018). Neural-network quantum state tomography. *Nature Physics, 14*(5), 447–450.
50. Tu, C., Zhang, W., Liu, Z., & Sun, M. (2016, July). Max-margin deepwalk: Discriminative learning of network representation. In *IJCAI* (Vol. 2016, pp. 3889–3895).
51. Uvarov, A., Kardashin, A., & Biamonte, J. (2019). Machine learning phase transitions with a quantum processor, 1, 1–32. *arXiv preprint arXiv:1906.10155.*
52. Van Nieuwenburg, E. P., Liu, Y. H., & Huber, S. D. (2017). Learning phase transitions by confusion. *Nature Physics, 13*(5), 435–439.

53. Venderley, J., Khemani, V., & Kim, E. A. (2018). Machine learning out-of-equilibrium phases of matter. *Physical review letters*, *120*(25), 257204.
54. Wang, L. (2016). Discovering phase transitions with unsupervised learning. *Physical Review B*, *94*(19), 195105.
55. Weiss, D., Alberti, C., Collins, M., & Petrov, S. (2015). Structured training for neural network transition-based parsing, 1, 1–30. *arXiv preprint arXiv:1506.06158*.
56. Wetzel, S. J., & Scherzer, M. (2017). Machine learning of explicit order parameters: From the Ising model to SU (2) lattice gauge theory. *Physical Review B*, *96*(18), 184410.
57. White, S. R. (2002). Numerical canonical transformation approach to quantum many-body problems. *The Journal of chemical physics*, *117*(16), 7472–7482.
58. Yang, C., Liu, Z., Zhao, D., Sun, M., & Chang, E. (2015, June). Network representation learning with rich text information. In *Twenty-Fourth International Joint Conference on Artificial Intelligence, 1, 1–30.*
59. Zdeborová, L. (2017). Machine learning: New tool in the box. *Nature Physics*, *13*(5), 420–421.

Chapter 8

Modern Applications of Machine Learning

CONTENTS

8.1. INTRODUCTION

A cognitive system attempts to perceive the approach of its environment with the help of a *model* where a model is a simplified analysis of this environment (Nguyen, 2010; Haroon, 2017). The strategy of building up such a model is known as *inductive learning*. Furthermore, a cognitive system is competent to formulate its experience by setting up new structures named *patterns*. Using a dataset, the building of models and patterns by a cognitive system is titled as *machine learning*. Two classifications of machine learning are given below: (Tzanis et al., 2006; Chomal & Saini, 2014):

- supervised learning;
- unsupervised learning.

A model depicts the entire set of data. As for a given value in the function's domain, the model can be utilized as well to predict the outcomes of a function (or *target function*), therefore it is marked as a *predictive model*. Additionally, some qualitative data information is provided by a model. Whereas a pattern specifies only a part of data and is symbolized as an *informative pattern* (Jaspree, 2011; Li, 2012).

8.1.1. Supervised Learning

Supervised learning is also called *learning from examples*. In this kind of learning, a function or a concept (i.e., description of a model) has been learned by the cognitive system. A set of instances are provided to the system specifically (Caruana & Niculescu-Mizil, 2006; Zhu & Goldberg, 2009). For every instance, the outcome of the target function is available as well. Based on the result of the function, the description of the model has been discovered by the system. For evaluation, a model builds from a subset of the data known as the *training set*, whereas the leftover data, known as the *test set*, are utilized to evaluate the designed model.

Two learning tasks are identified in supervised machine learning, i.e., (1) *classification* and (2) *regression*. The classification relates to the building of prediction models considering the functions having a discrete range. In contrast, regression deals with the building of prediction models considering the functions having continuous range. The following are the most generally known supervised machine learning methods (Chapelle et al., 2009; Kingma et al., 2014):

- *Concept Learning*. The instances are given to the cognitive system. Where a concept (or class) contains (positive instances) or do not have (negative instances). After that, the system is carried out to generate a generalized description of the class. So that based on this description, the decision can be taken for forthcoming instances.

- *Classification or Decision Tree Induction.* A most popular method, Classification or decision tree induction, is consumed for the approximation of discrete target functions. The training data is graphically represented in the form of tree structures constructed by these methods. They are easily interpreted. This is the main advantage of decision trees. The "*if-then*" rules can also be used to represent decision trees.
- *Rule Learning.* Rule learning comprises of induction of "*if-then*" rules, known as *classification rules*. For the approximation of discrete target functions, Classification rules are utilized.
- *Instance-Based Learning.* The data in raw form are stored in instance-based learning. When the system is executed for a new instance, it investigates the relationship of that instance with each of the stored instances for the decision of a new instance. As the learning process is delayed until a new instance appears. Therefore, this kind of learning is also named as *lazy learning.*
- *Bayesian Learning.* Bayes theorem is the base of Bayesian learning. It consists of methods that employ probabilities. The initial probabilities can incorporate into existing knowledge.
- *Linear Regression.* It is a method for depicting a target function along with a linear combination of several alternate variables. The range of the target function should be a continuous interval.
- *Neural Networks.* Classification and regression can be evaluated by neural networks. The function of neural networks is based on biological patterns and numerous processes that are used to imitate the human brain's activity.
- *Support Vector Machines (SVMs).* The base of SVMs is neural networks and statistical learning theory. SVMs are very famous in regression and classification jobs. Normally it gives very good prediction accuracy (Grandvalet & Bengio, 2005).

8.1.2. Unsupervised Learning

Unsupervised learning is called *learning from observation* as well. In this kind of learning, patterns based only on common attributes of instance have to be discovered by the system (i.e., clusters or associations) without knowing the quantity or even the presence of any patterns. Following are the main unsupervised machine learning methods (Barlow, 1989; Figueiredo & Jain, 2002):

- *Association Rule Mining.* Association rule mining came out in 1993. There is a lot of contribution to it from the research area of databases (Agrawal et al., 1993; Caron et al., 2018). It was proposed as a market basket analysis method. Association rules are the implications in the format $A \Rightarrow B$. This rule is interpreted as when item A show up in a basket, then item B will also show up in the identical basket (Ghahramani, 2003; Zhou et al., 2017).
- *Sequential Pattern Mining.* Sequential pattern mining involves learning from the ordered data. The temporal order is normally considered. There are also a lot of contributions to it from the research area of databases. It is considered to be an extension of association rules mining (Agrawal & Srikant, 1995).
- *Clustering.* The process of discovering clusters of instances such that instances belong to the same cluster are as similar as possible, whereas the instances belong to separate clusters are as dissimilar as possible is known as clustering (Hofmann, 2001; Le, 2013).

8.2. APPLICATIONS

Machine learning has been widely practiced in numerous application domains. Some of the well-known applications incorporate credit risk analysis, medical diagnosis, market segmentation, customer profiling, fraud detection, targeted marketing, and retail management. Since past years there is an availability of latest data and consequently, new domains have arisen where machine learning can be applied because of numerous technological advances and research exertion, e.g., the completion of the Human Genome Project (http://www.ornl.gov/sci/techresources/Human_Genome/home.shtml) and the growth of Web. Some of these contemporary applications are learning from text, learning from biological sequences, and learning in complex environments, i.e., Web. These contemporary applications of machine learning are given below (Koch et al., 2015). For today's technology, machine learning is a buzzword. It is expanding very speedily day by day. Machine learning is using in our routine life without even aware of it, i.e., Google assistant, Google Maps, Alexa, etc. The remarkable trending real-world applications of machine learning are given in Figure 8.1.

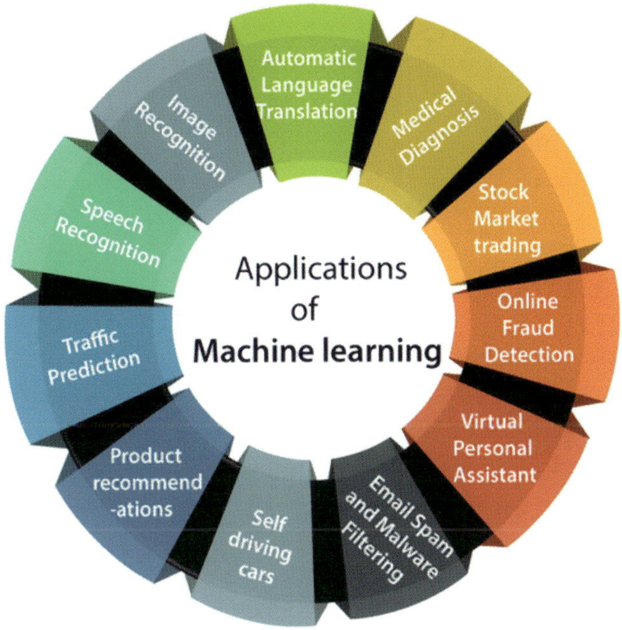

Figure 8.1. Utilization of machine learning.

[Source: https://www.javatpoint.com/applications-of-machine-learning.]

8.2.1. Image Recognition

One of the most prevalent applications of machine learning is image recognition. It is applied to identify persons, objects, digital images, places, etc. Automatic friend tagging suggestion is the famous use case of image recognition and face detection (Zoph et al., 2018):

Auto friend tagging suggestion is a feature provided by Facebook. Whenever a photo is uploaded with Facebook friends, then a tagging suggestion with name are automatically appears. Machine learning's face detection and recognition algorithm is the technology that supports this (Ayat et al., 2001; Svyrydov et al., 2018).

The Facebook project is known as "Deep Face" is subjected to personal identification and face recognition in the image. It is based on this project.

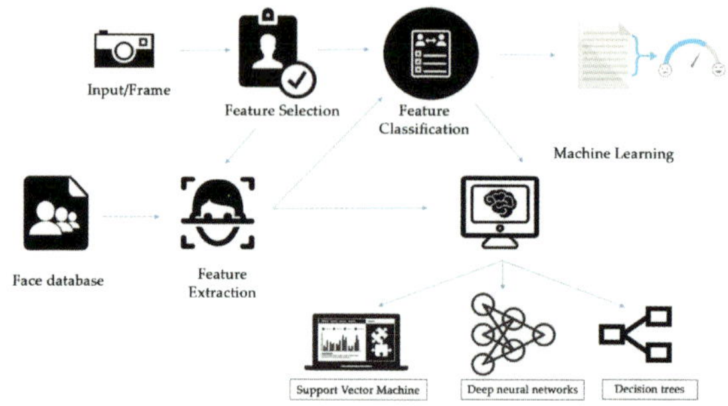

Figure 8.2. Face recognition with the help of machine learning.

[Source: https://www.researchgate.net/figure/Face-detection-and-emotion-recognition-using-machine-learning_fig1_322897107.]

8.2.2. Speech Recognition

An option of "Search by voice" is given while using Google. It is a rampant application of machine learning which is covered by speech recognition (Deng & Li, 2013; Qin et al., 2019).

A process of transforming voice instructions into text is known as speech recognition. It is also named as 'Computer speech recognition' or 'Speech to text.' Nowadays, numerous applications of speech recognition extensively use machine learning algorithms. Speech recognition technology is used by Google assistant, Cortana, Alexa, and Siri to pursue voice instructions (Skowronski & Harris, 2006; Amodei et al., 2016).

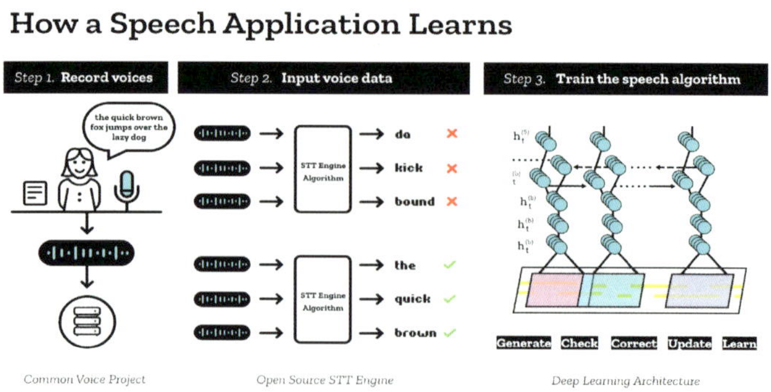

Figure 8.3. Voice recognition with the help of machine learning.

[Source: https://deepai.org/machine-learning-glossary-and-terms/automatic-speech-recognition.]

8.2.3. Traffic Prediction

If a new place is wanted to be visited, help from Google Maps is taken. Google Maps display accurate and shortest path and predicts traffic situation (Zhang et al., 2015; 2017).

It predicts the traffic situations like either traffic is cleared, heavily congested, or slow-moving. There are two ways to predict traffic conditions that are as follows:

- **Real-Time location** of the vehicle from sensors and Google Map app;
- **Average time has taken** on previous days at the same time.

The users of Google Maps are cooperating for the betterment of this app. The information is taken from the user and forwarded to its database to make improvements in the performance (Ide et al., 2015).

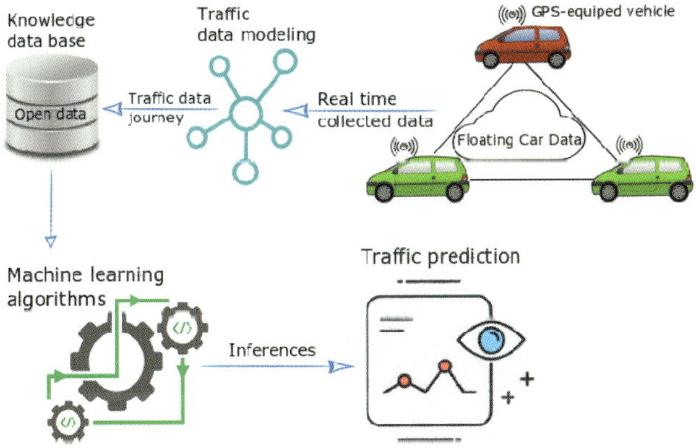

Figure 8.4. Traffic prediction demonstration through machine learning.

[Source:https://www.semanticscholar.org/paper/A-Novel-Traffic-Prediction-System-based-on-Floating-Mzibri-Maach/abe5fe88a8d-8f85725914e2e97e26d89a19812ef.]

8.2.4. Product Recommendations

For product recommendation to users, machine learning is broadly utilized by many entertainment and e-commerce enterprise, e.g., Netflix, Amazon, etc. Whenever some product is searched on Amazon, then due to machine learning an advertisement for the same product is showing during internet browsing on the same internet browser (Lee et al., 2012; Grbovic et al.,

2015). Because of different machine learning algorithms, Google figures out the interest of the user and advice the product as per user interest. Similarly, on the usage of Netflix, some recommendations for entertainment movies, series, etc., are found through machine learning.

8.2.5. Self-Driving Cars

Self-driving cars are one of the extremely astonishing applications of machine learning. There is a significant role of machine learning in self-driving cars. A famous car manufacturing company, Tesla, is working on a self-driving car. To detect objects and people during driving, the unsupervised learning method is used by them to train the car models (Maqueda et al., 2018; Stilgoe, 2018).

8.2.6. Email Spam and Malware Filtering

On receiving a new email, important, normal, and spam emails are automatically filtered. Important mails and spam emails are always received in an inbox with an important symbol and spam box respectively. Machine learning technology is behind this. Some spam filters used by Gmail are given below (Al-Jarrah et al., 2012; Dada et al., 2019):

- Header filter;
- Content filter;
- Rules-based filters;
- General blacklists filter; and
- Permission filters.

For malware detection and email spam filtering, machine learning algorithms, i.e., Decision tree, Naïve Bayes classifier, and Multi-Layer Perceptron are used.

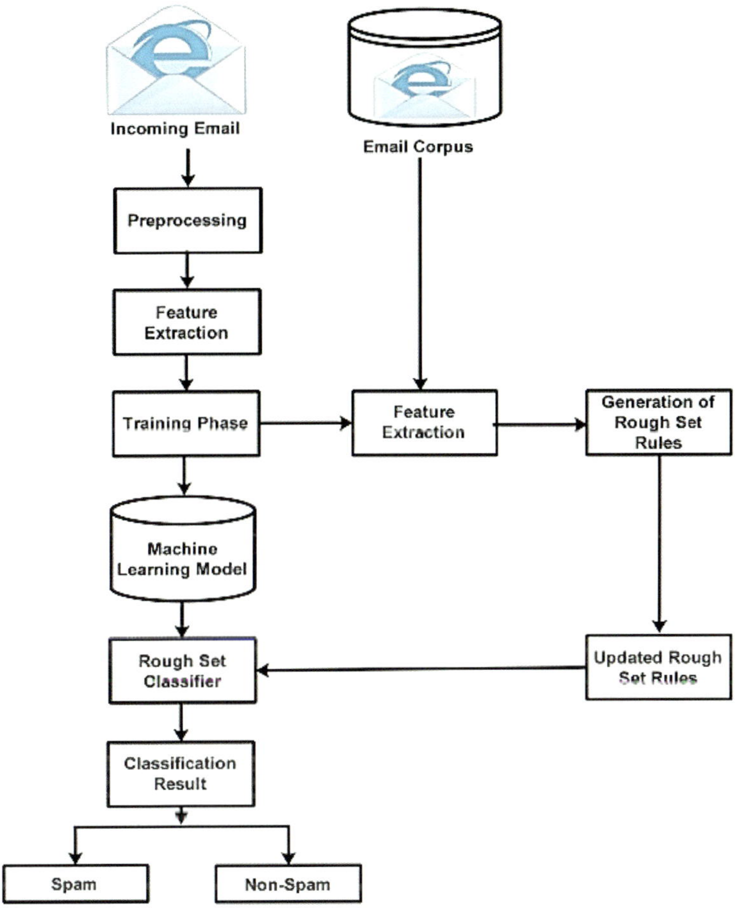

Figure 8.5. Email and Spam filter model via machine learning.

[Source:https://www.sciencedirect.com/science/article/pii/S2405844018353404.]

8.2.7. Virtual Personal Assistant

There are numerous virtual personal assistants, e.g., Alexa, Google Assistant, Siri, and Cortana. As it is cleared from its name, they assist to find out information with voice instruction (Gayathri et al., 2020). Using voice instructions, these assistants can help in diverse ways, e.g., call someone, Play music, Scheduling an appointment, Open an email, etc.

Machine learning algorithms are used as an important part of these virtual assistants.

The voice instruction is recorded, sent it over the server on a cloud, and decoded with the help of ML algorithms and acted correspondingly by these assistants (Kepuska & Bohouta, 2018).

8.2.8. Online Fraud Detection

Detection of fraud transactions (machine learning) makes our online transaction secure and safe. On performing any online transaction, a fraudulent transaction can be performed by numerous ways in between a transaction such as stolen money, fake accounts, and fake ids. Feed-Forward Neural network helps in the detection of a fraud or a genuine transaction (Awoyemi et al., 2017).

For every genuine transaction, the outcome is changed into some hash values, and then these values turn into the input for a later iteration. There is a particular pattern for every genuine transaction. A fraud transaction is detected when this pattern changed itself. Such that, machine learning makes online transactions safer (Adewumi & Akinyelu, 2017).

8.2.9. Stock Market trading

There is a broad application of machine learning in stock market trading. In the stock market, shares are at risk of ups and downs. So, to predict stock market trends, long short term memory neural network is used (Shen et al., 2012; Dash & Dash, 2016).

8.2.10. Medical Diagnosis

Machine learning plays an important role in disease diagnoses in medical science. With machine learning, there is a rapid growth in medical technology that makes us able to construct 3D models. The exact position of lesions can be estimated by these models in the brain.

Brain-related diseases especially brain tumors can be easily diagnosed with the help of it (Kononenko, 2001; Foster et al., 2014).

8.2.11. Automatic Language Translation

In the present day, when a new place is visited, then unfamiliarity about its language is not a problem anymore. Machine learning solves this problem by transforming the text into the languages that we can understand. This feature is provided by Google's GNMT (Google Neural Machine Translation). It is a Neural Machine Learning that converts the text into our known language. It is known as an automatic translation (Corston-Oliver et al., 2001).

Automatic translation works on the technology of a sequence to the sequence learning algorithm. Image recognition used it and converts the text from one language to another (Sagiroglu et al., 2007).

8.3. LEARNING FROM BIOLOGICAL SEQUENCES

In the past decades, the swift development of computer science has been closely pursued by similar rapid progress in molecular biology. No doubt, a rise in gathering and analysis of biological data has been given by the use of computational tools. It creates one of the hottest areas of research known as *bioinformatics*. An abundant fraction of biological data is represented by biological sequences that are analyzed with the help of the use of computational tools. The key reasons for the urgent need methods are the enormous range of the sequences and various possible features. They support the adequate analysis of such data and articulate accurate and reliable information to the domain specialist. The biologists are provided with a large set of gadgets for the examination of these data by the field of machine learning.

Various machine learning strategies have been presented to carry out the identification of explicit biological sequence segments. The generally known involves decision trees, Bayesian classifiers, Support Vector Machines, and neural networks (Hirsh & Noordewier, 1994; Ma & Wang, 1999). Sequence recognition algorithms display performance tradeoffs between reducing selectivity (ability to remove false positives) and growing sensitivity (ability to find out true positives). Nevertheless, it is stated by Li et al. (2003) that this kind of recognition problems cannot be solved directly by conventional machine learning methods. Therefore, the existing techniques are required to be adapted to this type of problem. Feature generation and feature selection are used to overcome this problem (Houle et al., 2000). Clustering algorithms are the alternate machine learning application that gathers structurally related biological sequences (Zien et al., 2000).

8.3.1. Prediction of Translation Initiation Sites

Proteins are the primary functional and structural molecules of an organism's cell. The genetic material of the organism encodes the information related to the synthesis of each protein. *DNA* is the genetic material of all living organisms. *RNA* is another molecule that plays a vital role in protein synthesis. RNA and DNA belong to a class of molecules known as *nucleic acids*. Both nucleic acids and proteins are sequences of smaller molecules, *nucleotides,* and *amino acids* respectively. A string of different symbols is a sequence. There are five nucleotides and twenty amino acids. Letters A, C, G, T, U characterize every nucleotide. DNA may consist of a consolidation of A, C, G, and T. U appears in RNA instead of T. By the given process, proteins are synthesized.

- *Transcription:* DNA is utilized as a template for the synthesis of RNA.
- *Translation*: Later RNA is utilized as a template for the synthesis of a protein molecule.

Translation occurs by an organelle named *ribosome*. Ribosome scans the mRNA sequence, which reads *codons*, or *triplets* of nucleotides and then "translates" them into amino acids. So, a sequence of $3n$ nucleotides encodes a protein consisting of n amino acids. Since an alphabet of four nucleotides forms 64 different triplets, and there is 20 total number of amino acids. More than one codon encodes some amino acids. Furthermore, as a translation initiation codon, triplet AUG is also used. Amino acid methionine is encoded by it. At last, three stop codons (UAG, UAA, and UGA) are applied for the termination of translation.

Normally, translation begins at the AUG codon closest to the outset of the RNA sequence. Nevertheless, it is not always happening. Because some escape strategies present that allow the inception of translation successive to AUG codons. On a given sequence, recognition of the Translation Initiation Site (TIS) becomes too complicated because of these strategies. After the translation starts, the ribosome progress with a molecule of RNA, towards the edge of the sequence and reads the upcoming codon. This mechanism is recurring until the ribosome attains stop codon. For every codon read, the proper amino acid is carried to the protein synthesis site and is merged to the protein chain, which by this technique is lengthened.

For more insights into the mechanism of translation, recognition of TIS is vital. It is considered as one of the most integral problems in molecular biology. It demands the production of classification models, so that valid TISs are distinguished from a set of false ones reliably and correctly (Kozak, 1987; Zeng et al., 2002).

Though, to overcome this problem, many methods have been suggested. But still, there is a tremendous capacity for the growth of their accuracy. In

Tzanis et al. (2005), the problem of the prediction of TISs in DNA sequences is dealt with by applying machine learning methods. To construct more accurate models, an abundant amount of features and various algorithms are used. There is direct extraction of few features from raw sequences, regarding the presence of nucleotides at every position of the sequence; on the other side most of the features are generated. Besides those features that are earlier used in other papers (Zeng et al., 2002), the use of some new features is generated and presented, i.e., *up-down_x*, *down_pos_k_x*, *up_pos_k_x* discussed in Table 1. It is demonstrated that the prediction models' accuracy is improved by the consolidation of these features. An extension is presented in Tzanis and Vlahavas (2006), where there was a corporation of a phase of organizing the sequences under the benchmark based on the length of the sequence. Additionally, amino acid patterns instead of nucleotide patterns were used being features. Ultimately, multiple classifier systems were erected. A real-world dataset is used in our experiments which consist of processed DNA sequences gathered from vertebrate organisms (Pedersen & Nielsen, 1997).

The approach that is pursued in Tzanis and Vlahavas (2006) to make up multiple classifier systems for the sake of prediction of TISs in genomic sequences is given in the following lines. Our approach comprises many steps. The following given lines depict each of these steps in detail.

- Step 1: All sequences are looked over and all nominee TIS is identified.
- Step 2: The nominee TISs identified in step 1 are grouped under sequence length after the AUG codon (downstream) and before the AUG codon (upstream). In this manner, several smaller datasets break down from the original dataset of candidate TISs. In this study, the original dataset is divided into 4 smaller datasets.
- Step 3: A calculation is performed for every candidate TISs to find out the value of the number of features. Table 8.1 enlists more insights into these features.
- Step 4: According to the significance of accuracy of classification, features are figured out from the cases of every group. In this study, information gain measure is used.
- Step 5: There is a selection of most participated features and a classifier is set up for every data subsets.

Table 8.1. Features Exploited in this Study (Tzanis & Vlahavas, 2006)

Feature	**Description**
up-down_*x*	Calculate the difference between the count of occurrences regarding amino acid *x* in the downstream region and the count of occurrences regarding amino acid *x* in the upstream region
up_*x*	Calculates the count of occurrences regarding amino acid *x* in the upstream region
down_*x*	Calculates the count of occurrences concerning amino acid *x* in the downstream region
down_pos_*k*_*x*	Calculates the count of occurrences concerning nucleotide *x* in the kth location of downstream codons (k ∈{1, 2, 3})
up_pos_*k*_*x*	Measures the count of occurrences regarding nucleotide *x* in the kth location of upstream codons (k ∈{1, 2, 3})
down_+1_G	A Boolean feature particularly true if there is a G nucleotide in initial position after the AUG codon, under Kozak's pattern (GCC[**AG**]CCaug**G**)
up_-3_[AG]	A Boolean feature particularly true if there is an A or a G nucleotide three spots back the AUG codon, under Kozak's pattern (GCC[**AG**]CCaug**G**)
down_stop	A Boolean feature particularly true if there is an in-frame downstream stop codon
up_AUG	A Boolean feature particularly true if there is an in-frame upstream AUG codon

Eventually, one of the groups assigns a new case a new candidate ATG under its upstream's length, and the length of the downstream region and corresponding classifier classifies it.

Numerous classification algorithms have tested for our approach (e.g., Naïve Bayes, C4.5, RIPPER, PART) and produced promising classification accuracy as compared to other approaches.

8.4. LEARNING FROM EMAIL DATA

From the prior years, massive popularity is given to email. Messages are sent and received daily; friends and partners get in touch and exchange information and files. Unfortunately, over the previous years, the paradox of email overload has developed becoming a financial concern for companies and intimate headaches for users. This section will discuss the contribution of machine learning for the solution to this situation (Katakis et al., 2006).

8.4.1. Automatic Answering

The email centers including "call centers" are normally maintained by large companies having employees that are dedicated to answering incoming messages. Usually, those messages arise from company partners and clients. And many of the times identical queries and problems are addressed by those messages. Personalized software or email centers are built as an effort for automatic email answering that will be competent to examine an incoming message and then suggest or even transmit an appropriate answer. Recently, this attempt has been made (Busemann et al., 2000; Bickel & Scheffer, 2004).

8.4.2. Automatic Mail Organization into Folders

A trend to find means of managing emails and organizing archives more efficiently came into users with the hike of email usage. Separate folders are used by many of the users to organize incoming messages. Topic-oriented, people-specific, or group-of-people-specific folders can be created. Topics-oriented such that 'personal,' 'funny,' and 'work,' people-specific such that 'Mary' and 'John,' or group-of-people-specific such that 'family,' 'friends,' and 'colleagues.' Messages are also being archived concerning its importance, and therefore folders maintained such that 'for future reference,' 'urgent,' 'spam' etc. Some formal *rules* are created manually by users to classify their email to achieve this goal.

Machine learning observes previous and recent classifications fashioned by the user to attempt an automatic classification of an incoming email (e.g., taking a recent classification as a model or studying already present folders). Hence, there is no need to create rules by the user himself. Additionally, the classification of the message can also be possible by machine learning algorithms; the content of messages is taken under consideration by searching for particular keywords. With the combination of linguistic and statistical techniques, this is commonly accomplished. It is exceptionally helpful for users, but still, some thoughts cannot be expressed easily with a sequence of keywords, e.g., 'interesting messages' or 'messages concerning my work' or 'messages that I have to answer today.' Besides, these thoughts may shift with time. For example, thought of 'interesting message.' With the observation of prior examples and by following the drift of thoughts using user feedback, machine learning algorithms can be learned silently for the classification of new messages (Clark et al., 2003; Eide et al., 2004; Klimt & Yang, 2004).

8.4.3. Email and Thread Summarization

A particular class of email users receives the bulk of messages daily. These messages can be appointment arrangements, business decision-making messages coming from colleagues, newsletters, etc. To avoid reading all these indicated messages rather than read-only necessary and most important parts, and then to decide either message require immediate attention would be remarkably convenient. In short, interesting content (e.g., from the newsletter) can be found out, and then the interesting full text can be read-only. For summarization, to set up trainable tools, data mining techniques are also probed (Muresan et al., 2001; Liu et al., 2003).

8.4.4. Spam Filtering

The identification and placement of unwelcome commercial emails (spam) from the user's mailbox (email stream) is the primary aim of spam filtering. In the past, spam emails have started as little discomfort. But from the last five years, it has become a considerable industry problem. A large number of spams are not only the reason of bandwidth problems (ultimately cause financial problems) but also consume the priceless time of email users in the separation and removal of many unwanted messages daily. Additionally, some of the spam messages contain pornographic content that is disproportionate for children. Various machine learning classifiers, i.e., Stacking Classifiers, Support Vector Machines, Naïve Bayes, have been evaluated in the bibliography and few of them have verified to be notably accurate (Sahami et al., 1998; Drucker et al., 1999; Sakkis et al., 2003).

8.4.5. Dynamic Feature Space and Incremental Feature Selection for Email Classification

The peculiar concern for Data Mining and Machine Learning community is email classification problems, chiefly because they offer and merge many significant difficulties. They incorporate unstructured, streaming, high dimensional, and, in many situations, concept drifting data. Another critical eccentricity of an email (in the general streaming text) that is not sufficiently explained in the related literature is the initial unavailability of the feature space. The computationally undemanding method is presented in this section that deals with this problem (Katakis et al., 2005). Two components are used in combination in our approach: (1) a method of an incremental feature ranking, and (2) an incremental learning algorithm that can recognize as a subset of the features while prediction. Filters are the feature selection methods generally utilized for text classification. They figured out the predictive power of every feature and choose the N best. Every word is evaluated by such methods based on cumulative statistics means the number of times each word turns out in each different group of documents. This interprets such methods intrinsically incremental: On the arrival of a new labeled document, there is a revision of statistics and without the demand for re-processing of previous data, evaluation can be instantly calculated. New words can also be managed by these methods by adding them in a dictionary (Vocabulary) and starting their statistics. Thus, such kinds of methods can instantiate the first component of our approach, i.e., the x^2 statistic, mutual information, or information gain.

The certain words will be unavoidably resulted by the addition and incremental re-evaluation of words being demoted from/ promoted to top N-words. A problem is raised that needs the second constituent of the presented approach: to classify a new case, the learning algorithm considers various features over time. The proper attention has not been given to this problem earlier. Those learning algorithms are known as *feature-based* algorithms that learn from the new subset of features, the same as instance-based algorithms that learn from the new instance. Naive Bayes (NB) is one of the intrinsically feature-based algorithms in which there is independent participation of each feature towards the prediction of a class. Hence, to instantiate the second constituent of our approach, it can be easily enlarged. Especially, extra parameters (designated the subset of the selected features) should be provided for the classification of a new instance using NB. It will then only view the computed probabilities of the previously mentioned subset. An algorithm for the sake of an incremental update of our approach is shown in Figure 8.6.

```
input  : Document, DocClass, Classes, Vocabulary
output : Classifier, Vocabulary, WordStats, Evaluation
begin
    foreach Word ∈ Document do
        if Word ∉ Vocabulary then
            AddWord(Word, Vocabulary)
            foreach Class ∈ Classes do
                WordStats [Word][Class][1] ← 0
                WordStats [Word][Class][0] ← 0
    foreach Word ∈ Vocabulary do
        if Word ∈ Document then
            WordStats [Word][DocClass][1] ← WordStats [Word][DocClass][1] + 1
        else
            WordStats [Word][DocClass][0] ← WordStats [Word][DocClass][0] + 1
    foreach Word ∈ Vocabulary do
        Evaluation ← EvaluateFeature(Word, WordStats)
    Classifier ← UpdateClassifier(Document, DocClass)
end
```

Figure 8.6. Algorithm for the sake of incremental updates.

[Source:https://www.researchgate.net/publication/228340464_Modern_Applications_of_Machine_Learning.]

8.5. FOCUSED CRAWLING USING REINFORCEMENT LEARNING

Any type of information that is required can be found at the World Wide Web. It is considered as the largest library. The web needs an indexing structure as compared to other libraries, which makes the users able to retrieve the required information. This problem is overwhelmed with the development of search engines that attempts to arrange and index web pages into pervasive catalogs. As many webpages as possible are gathered by search engines to encompass the majority of thematic subjects. The programs named '*crawlers*' are required to use in this process.

There is a simple structure of these programs. Following the included hyperlinks, they crawl the web initiated from a set of seed pages. Graph-based methods are used by initial approaches, e.g., breadth-first search and depth-first search. Crawling used in the above implementation is the resource and time-consuming process because of the stormy growth of the web. Furthermore, a group of users that have an interest in particular topics demanded correct, specific, and more related information. These demands are not satisfied by search engines which led towards the building of domain-specific search engines (Johnson et al., 2003).

The relevant webpages are collected to a peculiar domain by domain-specific search engines along with their index. Focused crawlers are used to carry out this task that is the agents and used the graph of the Web to search documents or pages belonging to a particular topic. In addition to that focused crawlers create thematic web portals and maintain them with recent and updated information.

Various methodologies have been introduced for the prior problem. The crawler is established on link criteria in Cho et al., (1998). While the knowledge about domains to build topic hierarchies are exploited by the system in Chakrabarti et al., (1999). Alternate approaches follow the learning and evolutionary algorithms respectively (Rennie & McCallum, 1999; O'Meara & Patel, 2001; Grigoriadis & Paliouras, 2004).

8.5.1. Reinforcement Learning

In the problems of Reinforcement Learning (RL), agents learn through the trial-and-error approach with the dynamic environment (Pant & Menczer, 2002). In an RL activity, at each iteration agent perceive the state of the environment, s_t in S (S →a finite set of possible states), and choose an action $A(s_t)$ for execution, where $A(s_t)$ → finite set of possible actions in state s_t. A reward, r_{t+1} in R, is received by an agent, and the agent steps forward to a new state s_{t+1}. The goal of an agent is to boost the received cumulative reward over time. More particularly, the actions are selected by the agent that increase the expected discounted return:

$$R_t = r_{t+1} + \gamma r_{t+2} + \gamma^2 r_{t+3} + \ldots = \sum_{k=0}^{\infty} \gamma^k r_{t+k+1}$$

where γ is the discount factor which indicates the significance of future rewards. A *policy* π determines that the probability of executing an action α in state *s* is π(s,α). The *action-value function,* $Q^{\pi}(s, a)$ for any policy π, can be expressed as in state *s* the expected discounted return for taking action *a* and thenceforth following π. For all state-action pairs, optimal policy, π* inflates the action-value $Q^{\pi}(s, a)$. To learn the optimal policy, the optimal action-value function Q^* is learned by the agent that is described as the expected return of executing action α in state *s* and thenceforth following the optimal policy π*. For finding the optimal policy, the Q-learning algorithm is the most commonly used (Sutton & Barto, 2018).

8.5.2. Methodology

The following components must be defined to draw up the problem of focused crawling being an RL problem.

S, a set of states: The state is expressed as the crawler access the pages, as the environment sense comes up primarily by the pages fetched at any given time.

A, a set of actions: Actions are expressed as the classes of links from which the crawler has to make a selection on visiting a particular page.

r(s, a), a reward function: On page visit, a reward *r* is received by the crawler, s:

$$r = \begin{cases} 1 & if\ s = target\ page \\ 0 & otherwise \end{cases}$$

Several episodes are executed by the crawler in the training phase initiated from a seed set of pages. When the desired page is found or several predefined iterations are reached, the episode ends. When a page is visited by a crawler, the hyperlinks that the page holds are classified based on the page's textual information that they connect to. A class of links is selected by the crawler from where a URL is selected to follow. A reward is received by the crawler and transits to the afterward page. The ε-greedy action selection method is used to probe the state space where an action *a* is chosen according to the following mentioned rule:

$$a = \begin{cases} a\ random\ action\ with\ probability\ \varepsilon \\ \arg\max_{a'} Q(s,a')\ with\ probability\ 1-\varepsilon \end{cases}$$

At the crawling stage, the crawler begins with a set of seed pages, and a queue is maintained with a list of unvisited hyperlinks. There are extraction and classification of the hyperlinks from each page. In the training phase, the function that the crawler learned is used for the evaluation of each action. The link with the highest relevance score is selected by the crawler. The same process is repeated until the specific number of pages are visited or on an empty queue.

REFERENCES

1. Adewumi, A. O., & Akinyelu, A. A. (2017). A survey of machine-learning and nature-inspired based credit card fraud detection techniques. *International Journal of System Assurance Engineering and Management*, *8*(2), 937–953.
2. Agrawal, R., & Srikant, R. (1995, March). Mining sequential patterns. In *Proceedings of the Eleventh International Conference on Data Engineering* (Vol.1, pp. 3–14). IEEE.
3. Agrawal, R., Imieliński, T., & Swami, A. (1993, June). Mining association rules between sets of items in large databases. In *Proceedings of the 1993 ACM SIGMOD International Conference on Management of Data* (Vol.1, pp. 207–216).
4. Al-Jarrah, O., Khater, I., & Al-Duwairi, B. (2012). Identifying potentially useful email header features for email spam filtering. In *The Sixth International Conference on Digital Society (ICDS)* (Vol. 30, p. 140).
5. Amodei, D., Ananthanarayanan, S., Anubhai, R., Bai, J., Battenberg, E., Case, C., ... & Chen, J. (2016, June). Deep speech 2: End-to-end speech recognition in english and mandarin. In *International Conference on Machine Learning* (Vol.1, pp. 173–182).
6. Awoyemi, J. O., Adetunmbi, A. O., & Oluwadare, S. A. (2017, October). Credit card fraud detection using machine learning techniques: A comparative analysis. In *2017 International Conference on Computing Networking and Informatics (ICCNI)* (Vol.1, pp. 1–9). IEEE.
7. Ayat, N. E., Cheriet, M., Remaki, L., & Suen, C. Y. (2001, September). KMOD-A new support vector machine kernel with moderate decreasing for pattern recognition. Application to digit image recognition. In *Proceedings of Sixth International Conference on Document Analysis and Recognition* (Vol.1, pp. 1215–1219). IEEE.
8. Barlow, H. B. (1989). Unsupervised learning. *Neural Computation*, *1*(3), 295–311.
9. Bickel, S., & Scheffer, T. (2004, September). Learning from message pairs for automatic email answering. In *European Conference on Machine Learning* (Vol.1, pp. 87–98). Springer, Berlin, Heidelberg.
10. Busemann, S., Schmeier, S., & Arens, R. G. (2000). Message classification in the call center, 1, 1–33. *arXiv preprint cs/0003060.*

11. Caron, M., Bojanowski, P., Joulin, A., & Douze, M. (2018). Deep clustering for unsupervised learning of visual features. In *Proceedings of the European Conference on Computer Vision (ECCV)* (Vol.1, pp. 132–149).
12. Caruana, R., & Niculescu-Mizil, A. (2006, June). An empirical comparison of supervised learning algorithms. In *Proceedings of the 23rd International Conference on Machine Learning* (Vol.1, pp. 161–168).
13. Chakrabarti, S., Van den Berg, M., & Dom, B. (1999). Focused crawling: a new approach to topic-specific Web resource discovery. *Computer networks*, *31*(11–16), 1623–1640.
14. Chapelle, O., Scholkopf, B., & Zien, A. (2009). Semi-supervised learning (chapelle, o. et al., eds.; 2006)[book reviews]. *IEEE Transactions on Neural Networks*, *20*(3), 542–542.
15. Cho, J., Garcia-Molina, H., & Page, L. (1998). Efficient crawling through URL ordering, 1, 1–33.
16. Chomal, V. S., & Saini, J. R. (2014). Cataloguing most severe causes that lead software projects to fail. *International Journal on Recent and Innovation Trends in Computing and Communication*, *2*(5), 1143–1147.
17. Clark, J., Koprinska, I., & Poon, J. (2003, October). A neural network based approach to automated e-mail classification. In *Proceedings IEEE/WIC International Conference on Web Intelligence (WI 2003)* (Vol.1, pp. 702–705). IEEE.
18. Corston-Oliver, S., Gamon, M., & Brockett, C. (2001, July). A machine learning approach to the automatic evaluation of machine translation. In *Proceedings of the 39th Annual Meeting of the Association for Computational Linguistics* (Vol.1, pp. 148–155).
19. Dada, E. G., Bassi, J. S., Chiroma, H., Adetunmbi, A. O., & Ajibuwa, O. E. (2019). Machine learning for email spam filtering: review, approaches and open research problems. *Heliyon*, *5*(6), e01802.
20. Dash, R., & Dash, P. K. (2016). A hybrid stock trading framework integrating technical analysis with machine learning techniques. *The Journal of Finance and Data Science*, *2*(1), 42–57.
21. Deng, L., & Li, X. (2013). Machine learning paradigms for speech recognition: An overview. *IEEE Transactions on Audio, Speech, and Language Processing*, *21*(5), 1060–1089.

22. Drucker, H., Wu, D., & Vapnik, V. N. (1999). Support vector machines for spam categorization. *IEEE Transactions on Neural networks*, *10*(5), 1048–1054.
23. Eide, N. E., Blaafadt, A. N., Johansen, B. H. R., & Sandnes, F. E. (2004, November). DIGIMIMIR: A Tool for Rapid Situation Analysis of Helpdesk and Support Email. In *LISA* (Vol.1, pp. 21–32).
24. Figueiredo, M. A. T., & Jain, A. K. (2002). Unsupervised learning of finite mixture models. *IEEE Transactions on Pattern Analysis and Machine Intelligence*, *24*(3), 381–396.
25. Foster, K. R., Koprowski, R., & Skufca, J. D. (2014). Machine learning, medical diagnosis, and biomedical engineering research-commentary. *Biomedical engineering online*, *13*(1), 94.
26. Gayathri, M., Malathy, C., & Singh, S. (2020). MARK42: the secured personal assistant using biometric traits integrated with green IOT. *J Green Eng*, *10*(1), 255–267.
27. Ghahramani, Z. (2003, February). Unsupervised learning. In *Summer School on Machine Learning* (Vol.1, pp. 72–112). Springer, Berlin, Heidelberg.
28. Grandvalet, Y., & Bengio, Y. (2005). Semi-supervised learning by entropy minimization. In *Advances in Neural Information Processing Systems* (Vol.1, pp. 529–536).
29. Grbovic, M., Radosavljevic, V., Djuric, N., Bhamidipati, N., Savla, J., Bhagwan, V., & Sharp, D. (2015, August). E-commerce in your inbox: Product recommendations at scale. In *Proceedings of the 21th ACM SIGKDD international conference on knowledge discovery and data mining* (Vol.1, pp. 1809–1818).
30. Grigoriadis, A., & Paliouras, G. (2004, May). Focused crawling using temporal difference-learning. In *Hellenic Conference on Artificial Intelligence* (Vol.1, pp. 142–153). Springer, Berlin, Heidelberg.
31. Haroon, D. (2017). Clustering. In *Python Machine Learning Case Studies* (Vol.1, pp. 129–160). Apress, Berkeley, CA.
32. Hirsh, H., & Noordewier, M. (1994, March). Using background knowledge to improve inductive learning of DNA sequences. In *Proceedings of the Tenth Conference on Artificial Intelligence for Applications* (Vol.1, pp. 351–357). IEEE.
33. Hofmann, T. (2001). Unsupervised learning by probabilistic latent semantic analysis. *Machine learning*, *42*(1–2), 177–196.

34. Houle, J. L., Cadigan, W., Henry, S., Pinnamaneni, A., & Lundahl, S. (2000). Database mining in the human genome initiative. *Genome, 3*, 1–6.
35. Ide, C., Hadiji, F., Habel, L., Molina, A., Zaksek, T., Schreckenberg, M., ... & Wietfeld, C. (2015, September). LTE connectivity and vehicular traffic prediction based on machine learning approaches. In *2015 IEEE 82nd Vehicular Technology Conference (VTC2015-Fall)* (Vol.1, pp. 1–5). IEEE.
36. Jaspree, K. (2011). P. S., "A k-means Based Approach for Prediction of Level of Severity of Faults in Software System." In *Proceedings of International conference on Intelligent Computational Systems, 1, 1–33.*
37. Johnson, J., Tsioutsiouliklis, K., & Giles, C. L. (2003). Evolving strategies for focused web crawling. In *Proceedings of the 20th International Conference on Machine Learning (ICML-03)* (Vol.1, pp. 298–305).
38. Katakis, I., Tsoumakas, G., & Vlahavas, I. (2005, November). On the utility of incremental feature selection for the classification of textual data streams. In *Panhellenic Conference on Informatics* (Vol.1, pp. 338–348). Springer, Berlin, Heidelberg.
39. Katakis, I., Tsoumakas, G., & Vlahavas, I. (2006). Email mining: Emerging techniques for email management (to appear), Web Data Management Practices: Emerging Techniques and Technologies (Vol.1, pp. 1–22), Athena Vakali, George Pallis.
40. Kepuska, V., & Bohouta, G. (2018, January). Next-generation of virtual personal assistants (microsoft cortana, apple siri, amazon alexa and google home). In *2018 IEEE 8th Annual Computing and Communication Workshop and Conference (CCWC)* (pp. 99–103). IEEE.
41. Kingma, D. P., Mohamed, S., Rezende, D. J., & Welling, M. (2014). Semi-supervised learning with deep generative models. In *Advances in Neural Information Processing Systems* (Vol.1, pp. 3581–3589).
42. Klimt, B., & Yang, Y. (2004, September). The enron corpus: A new dataset for email classification research. In *European Conference on Machine Learning* (Vol. 1, pp. 217–226). Springer, Berlin, Heidelberg.
43. Koch, G., Zemel, R., & Salakhutdinov, R. (2015, July). Siamese neural networks for one-shot image recognition. In *ICML deep learning workshop* (Vol. 2, pp. 1–33).

44. Kononenko, I. (2001). Machine learning for medical diagnosis: history, state of the art and perspective. *Artificial Intelligence in Medicine, 23*(1), 89–109.
45. Kozak, M. (1987). An analysis of 5'-noncoding sequences from 699 vertebrate messenger RNAs. *Nucleic acids research, 15*(20), 8125–8148.
46. Le, Q. V. (2013, May). Building high-level features using large scale unsupervised learning. In *2013 IEEE International Conference on Acoustics, Speech, and Signal Processing* (Vol.1, pp. 8595–8598). IEEE.
47. Lee, Y. H., Hu, P. J. H., Cheng, T. H., & Hsieh, Y. F. (2012). A cost-sensitive technique for positive-example learning supporting content-based product recommendations in B-to-C e-commerce. *Decision Support Systems, 53*(1), 245–256.
48. Li, J., See-Kiong, N., & Wong, L. (2003, January). Bioinformatics adventures in database research. In *International Conference on Database Theory* (Vol.1, pp. 31 46). Springer, Berlin, Heidelberg.
49. Li, W. B. (2012). Study on supply chain partner selection based on support vector machine. In *Applied Mechanics and Materials* (Vol. 121, pp. 4779–4783). Trans Tech Publications Ltd.
50. Liu, J., Faltings, B., Zhong, N., Lu, R., & Nishida, T. (2003). IEEE/WIC International Conference on Intelligent Agent Technology, 1, 1–33.
51. Ma, Q., & Wang, J. T. (1999). Biological data mining using Bayesian neural networks: A case study. *International Journal on Artificial Intelligence Tools, 8*(04), 433–451.
52. Maqueda, A. I., Loquercio, A., Gallego, G., García, N., & Scaramuzza, D. (2018). Event-based vision meets deep learning on steering prediction for self-driving cars. In *Proceedings of the IEEE Conference on Computer Vision and Pattern Recognition* (Vol.1, pp. 5419–5427).
53. Muresan, S., Tzoukermann, E., & Klavans, J. L. (2001). Combining linguistic and machine learning techniques for email summarization. In *Proceedings of the ACL 2001 Workshop on Computational Natural Language Learning (ConLL), 1, 1–33.*
54. Nguyen, V. (2010). The Application of Machine Learning Methods in Software Verification and Validation (Vol.1, pp. 1–33). *MSc, Master of Science in Engineering, The University of Texas at Austin, Texas.*

55. O'Meara, T., & Patel, A. (2001). A topic-specific Web robot model based on restless bandits. *IEEE Internet Computing*, *5*(2), 27–35.
56. Pant, G., & Menczer, F. (2002). MySpiders: Evolve your own intelligent Web crawlers. *Autonomous agents and multi-agent systems*, *5*(2), 221–229.
57. Pedersen, A. G., & Nielsen, H. (1997, June). Neural network prediction of translation initiation sites in eukaryotes: perspectives for EST and genome analysis. In *Ismb* (Vol. 5, pp. 226–233).
58. Qin, Y., Carlini, N., Cottrell, G., Goodfellow, I., & Raffel, C. (2019, May). Imperceptible, robust, and targeted adversarial examples for automatic speech recognition. In *International Conference on Machine Learning* (Vol.1, pp. 5231–5240).
59. Rennie, J., & McCallum, A. (1999). Efficient web spidering with reinforcement learning. In *Proceedings of the International Conference on Machine Learning, 1, 1–33.*
60. Sagiroglu, S., Yavanoglu, U., & Guven, E. N. (2007, December). Web based machine learning for language identification and translation. In *Sixth International Conference on Machine Learning and Applications (ICMLA 2007)* (Vol.1, pp. 280–285). IEEE.
61. Sahami, M., Dumais, S., Heckerman, D., & Horvitz, E. (1998, July). A Bayesian approach to filtering junk e-mail. In *Learning for Text Categorization: Papers from the 1998 Workshop* (Vol. 62, pp. 98–105).
62. Sakkis, G., Androutsopoulos, I., Paliouras, G., Karkaletsis, V., Spyropoulos, C. D., & Stamatopoulos, P. (2003). A memory-based approach to anti-spam filtering for mailing lists. *Information retrieval*, *6*(1), 49–73.
63. Shen, S., Jiang, H., & Zhang, T. (2012). Stock market forecasting using machine learning algorithms. *Department of Electrical Engineering,* Stanford University, Stanford, CA, *1*, 1–5.
64. Skowronski, M. D., & Harris, J. G. (2006). Acoustic detection and classification of microchiroptera using machine learning: lessons learned from automatic speech recognition. *The Journal of the Acoustical Society of America*, *119*(3), 1817–1833.
65. Stilgoe, J. (2018). Machine learning, social learning and the governance of self-driving cars. *Social Studies of Science*, *48*(1), 25–56.
66. Sutton, R. S., & Barto, A. G. (2018). *Reinforcement learning: An introduction* (Vol.1, pp. 1–31). MIT Press.

67. Svyrydov, A., Kuchuk, H., & Tsiapa, O. (2018, May). Improving efficienty of image recognition process: Approach and case study. In *2018 IEEE 9th International Conference on Dependable Systems, Services and Technologies (DESSERT)* (Vol.1, pp. 593–597). IEEE.
68. Tzanis, G., & Vlahavas, I. (2006, May). Prediction of translation initiation sites using classifier selection. In *Hellenic Conference on Artificial Intelligence* (Vol.1, pp. 367–377). Springer, Berlin, Heidelberg.
69. Tzanis, G., Berberidis, C., Alexandridou, A., & Vlahavas, I. (2005, November). Improving the accuracy of classifiers for the prediction of translation initiation sites in genomic sequences. In *Panhellenic Conference on Informatics* (Vol. 1, pp. 426–436). Springer, Berlin, Heidelberg.
70. Tzanis, G., Katakis, I., Partalas, I., & Vlahavas, I. (2006, July). Modern applications of machine learning. In *Proceedings of the 1st Annual SEERC Doctoral Student Conference–DSC* (Vol. 1, No. 1, pp. 1–10).
71. Zeng, F., Yap, R. H., & Wong, L. (2002). Using feature generation and feature selection for accurate prediction of translation initiation sites. *Genome Informatics*, *13*, 192–200.
72. Zhang, S., Zhao, S., Yuan, M., Zeng, J., Yao, J., Lyu, M. R., & King, I. (2017, November). Traffic prediction based power saving in cellular networks: A machine learning method. In *Proceedings of the 25th ACM SIGSPATIAL International Conference on Advances in Geographic Information Systems* (Vol.1, pp. 1–10).
73. Zhang, Y., Ni, F., Feng, Z., Cui, S., & Ding, Z. (2015, July). Wavelet transform processing for cellular traffic prediction in machine learning networks. In *2015 IEEE China Summit and International Conference on Signal and Information Processing (ChinaSIP)* (Vol.1, pp. 458–462). IEEE.
74. Zhou, T., Brown, M., Snavely, N., & Lowe, D. G. (2017). Unsupervised learning of depth and ego-motion from video. In *Proceedings of the IEEE Conference on Computer Vision and Pattern Recognition* (Vol.1, pp. 1851–1858).
75. Zhu, X., & Goldberg, A. B. (2009). Introduction to semi-supervised learning. *Synthesis Lectures on Artificial Intelligence and Machine Learning*, *3*(1), 1–130.

76. Zien, A., Rätsch, G., Mika, S., Schölkopf, B., Lengauer, T., & Müller, K. R. (2000). Engineering support vector machine kernels that recognize translation initiation sites. *Bioinformatics*, *16*(9), 799–807.
77. Zoph, B., Vasudevan, V., Shlens, J., & Le, Q. V. (2018). Learning transferable architectures for scalable image recognition. In *Proceedings of the IEEE Conference on Computer Vision and Pattern Recognition* (Vol.1, pp. 8697–8710).

INDEX

U

V

W